ZeroMQ

Use ZeroMQ and learn how to apply different message patterns

Faruk Akgul

[PACKT] PUBLISHING open source community experience distilled

BIRMINGHAM - MUMBAI

ZeroMQ

First published: March 2013

Production Reference: 1140313

Published by Packt Publishing Ltd.
Livery Place
35 Livery Street
Birmingham B3 2PB, UK.

ISBN 978-1-78216-104-2

www.packtpub.com

Cover Image by Abhishek Pandey (abhishek.pandey1210@gmail.com)

Credits

Author
Faruk Akgul

Reviewers
Burak Arslan
David Greco
Kevin J. Rice

Acquisition Editor
Usha Iyer

Commissioning Editor
Harsha Bharwani

Technical Editor
Hardik Soni

Copy Editors
Insiya Morbiwala
Alfida Paiva
Laxmi Subramanian

Project Coordinator
Amigya Khurana

Proofreader
Maria Gould

Indexer
Monica Ajmera Mehta

Graphics
Valentina D'silva

Production Coordinator
Manu Joseph

Cover Work
Manu Joseph

About the Author

Faruk Akgul is a developer and an Emacs user who loves using open source software and frequently contributes to some open source projects. He specializes in Python but enjoys experiencing new programming languages as well. He likes to travel when he's not coding.

Thanks to Pieter Hintjens and all the contributors who helped in making this software available.

About the Reviewers

Burak Arslan is the technical lead at Arskom, a quarter-century-old Turkish Software and IT infrastructure services company focusing on bundling the latest products in Satellite Communications with its value-added solutions. His current job involves working as a full-stack web developer where he gets to design and implement the UX, frontend, and backend code, as well as administering the underlying IP network and hardware infrastructure, while also having an influence in managing customer relations and the future strategy and planning of his company. He has a BSc in Computer Engineering from Galatasaray University and an MSc in Computer Science from Sabanci University, both in Istanbul, Turkey.

David Greco is an experienced software architect with more than 20 years of working experience. After an initial period as a researcher in the field of high performance computing, he started working as a consultant in the professional services organizations of leading software companies such as BEA Systems, IONA, Progress, and FuseSource. As a consultant, he has mainly helped customers to design and develop complex distributed platforms and service-oriented architectures. Lately, he worked as a CTO for one of the most successful gambling and poker online companies in Italy.

David is now working as a CTO for a startup, Eligotech, developing a parallel business intelligence platform based on very popular big data technologies.

Kevin Rice has been involved in computing since the mid 80s, starting with a TI-99/4A and Commodore 64. His current skill set includes Unix, C/C++, PERL, Python, infosec, and network-monitoring related tools.

I'd like to thank Packt Publishing and Harsha Bharwani for allowing me the opportunity to review this book.

www.PacktPub.com

Support files, eBooks, discount offers and more

You might want to visit www.PacktPub.com for support files and downloads related to your book.

Did you know that Packt offers eBook versions of every book published, with PDF and ePub files available? You can upgrade to the eBook version at www.PacktPub.com and as a print book customer, you are entitled to a discount on the eBook copy. Get in touch with us at service@packtpub.com for more details.

At www.PacktPub.com, you can also read a collection of free technical articles, sign up for a range of free newsletters and receive exclusive discounts and offers on Packt books and eBooks.

PACKTLIB

http://PacktLib.PacktPub.com

Do you need instant solutions to your IT questions? PacktLib is Packt's online digital book library. Here, you can access, read and search across Packt's entire library of books.

Why Subscribe?

- Fully searchable across every book published by Packt
- Copy and paste, print and bookmark content
- On demand and accessible via web browser

Free Access for Packt account holders

If you have an account with Packt at www.PacktPub.com, you can use this to access PacktLib today and view nine entirely free books. Simply use your login credentials for immediate access.

Table of Contents

Preface

This book is an introductory guide to message queuing components and ZeroMQ. We will cover how you can apply patterns to your applications.

What this book covers

Chapter 1, Getting Started, explains what a message queuing system is, discusses the importance of message queuing, and introduces ZeroMQ to the reader. It also introduces the request-reply pattern with the "hello world" example and shows examples of how to handle string in C and version reporting in ZeroMQ.

Chapter 2, Introduction to Sockets, explores how to use ZeroMQ with sockets by providing example code.

Chapter 3, Using Socket Topology, goes beyond *Chapter 2, Introduction to Sockets*, and discusses the difference between ZeroMQ sockets and TCP sockets and shows how to use the topics covered in the previous chapters for real-world applications.

Chapter 4, Advanced Patterns, is a brief introduction to more advanced topics and discusses how to use patterns in ZeroMQ applications.

Appendix, contains bibliography and external links.

What you need for this book

To run the examples in the book the following software will be required:

- ZeroMQ v3.2, available at `http://www.zeromq.org/`
- CZMQ v1.3.1, available at `http://czmq.zeromq.org/`

- Microsoft Visual C++ (to build on Windows), available at `http://www.microsoft.com/visualstudio/eng/products/visual-studio-express-products`
- GCC v4.7.2, available at `http://gcc.gnu.org`
- The Libtool, Autoconf, and Automake tools to build on Unix

Who this book is for

This book is for developers who are interested in learning and implementing ZeroMQ for their applications. The reader needs to have basic C programming knowledge. Prior ZeroMQ experience is not expected.

Conventions

In this book, you will find a number of styles of text that distinguish between different kinds of information. Here are some examples of these styles, and an explanation of their meaning.

Code words in text are shown as follows: "You may use `czmq.h` which lets C developers to code their ZeroMQ applications easier and shorter."

A block of code is set as follows:

```
#include <string.h>
#include <stdio.h>
#include "zmq.h"


int main (int argc, char const *argv[]) {
  void* context = zmq_init(1);
  void* request = zmq_socket(context, ZMQ_REQ);
  printf("Connecting to server\n");
  zmq_connect(request, "tcp://localhost:4040");

  zmq_close(request);
  zmq_term(context);
  return 0;
}
```

When we wish to draw your attention to a particular part of a code block, the relevant lines or items are set in bold:

```
#include <string.h>
#include <stdio.h>
#include "zmq.h"

int main (int argc, char const *argv[]) {
  void* context = zmq_init(1);
  void* request = zmq_socket(context, ZMQ_REQ);
  printf("Connecting to server\n");
  zmq_connect(request, "tcp://localhost:4040");

  zmq_close(request);
  zmq_term(context);
  return 0;
}
```

Any command-line input or output is written as follows:

```
gcc -Wall -lzmq -o zero zero.c
```

When we say `zmq_socket(2)` we mean `zmq_socket` function takes two parameters. When we say `zmq_ctx_new()` we mean `zmq_ctx_new` function does not take any parameters.

New terms and **important words** are shown in bold.

> Warnings or important notes appear in a box like this.

> Tips and tricks appear like this.

Reader feedback

Feedback from our readers is always welcome. Let us know what you think about this book—what you liked or may have disliked. Reader feedback is important for us to develop titles that you really get the most out of.

To send us general feedback, simply send an e-mail to `feedback@packtpub.com`, and mention the book title via the subject of your message.

If there is a topic that you have expertise in and you are interested in either writing or contributing to a book, see our author guide on `www.packtpub.com/authors`.

Customer support

Now that you are the proud owner of a Packt book, we have a number of things to help you to get the most from your purchase.

Downloading the example code

You can download the example code files for all Packt books you have purchased from your account at `http://www.PacktPub.com`. If you purchased this book elsewhere, you can visit `http://www.PacktPub.com/support` and register to have the files e-mailed directly to you.

Errata

Although we have taken every care to ensure the accuracy of our content, mistakes do happen. If you find a mistake in one of our books—maybe a mistake in the text or the code—we would be grateful if you would report this to us. By doing so, you can save other readers from frustration and help us improve subsequent versions of this book. If you find any errata, please report them by visiting `http://www.packtpub.com/support`, selecting your book, clicking on the **errata submission form** link, and entering the details of your errata. Once your errata are verified, your submission will be accepted and the errata will be uploaded on our website, or added to any list of existing errata, under the Errata section of that title. Any existing errata can be viewed by selecting your title from `http://www.packtpub.com/support`.

Piracy

Piracy of copyright material on the Internet is an ongoing problem across all media. At Packt, we take the protection of our copyright and licenses very seriously. If you come across any illegal copies of our works, in any form, on the Internet, please provide us with the location address or website name immediately so that we can pursue a remedy.

Please contact us at `copyright@packtpub.com` with a link to the suspected pirated material.

We appreciate your help in protecting our authors, and our ability to bring you valuable content.

Questions

You can contact us at `questions@packtpub.com` if you are having a problem with any aspect of the book, and we will do our best to address it.

1
Getting Started

Welcome to ZeroMQ! This chapter is an introduction to ZeroMQ and gives the reader a general idea of what a message queuing system is and most importantly what ZeroMQ is. In this chapter we will learn about the following topics:

- An overview of what a message queue is
- Why use ZeroMQ and what makes it different from other message queuing technologies
- Basic client/server architecture
- Introducing the first pattern, request-reply
- How we can handle strings in C
- Detecting the installed ZeroMQ version

The beginning

Humans are social and will always socially interact with each other for as long as they exist. Programs are no different. A program has to communicate with another program since we are living in a connected world. We have UDP, TCP, HTTP, IPX, WebSocket, and other relevant protocols to connect applications.

However, such low-level approaches make things harder and we need something easier and faster. High-level abstractions sacrifice speed and flexibility whereas directly dealing with low-level details is not easy to master and use. That is where ZeroMQ shows up as the savior, giving us the usability features of high-level techniques with the speed of low-level approaches.

Before we start digging into ZeroMQ, let's first have a brief introduction on the general concept of message queues.

The message queue

A message queue, or technically a FIFO (First In First Out) queue is a fundamental and well-studied data structure. There are different queue implementations such as priority queues or double-ended queues that have different features, but the general idea is that the data is added in a queue and fetched when the data or the caller is ready.

Imagine we are using a basic in-memory queue. In case of an issue, such as power outage or a hardware failure, the entire queue could be lost. Hence, another program that expects to receive a message will not receive any messages.

However, adopting a message queue guarantees that messages will be delivered to the destination no matter what happens. Message queuing enables asynchronous communication between loosely-coupled components and also provides solid queuing consistency. In case of insufficient resources, which prevent you from immediately processing the data that is sent, you can queue them up in the message queue server that would store the data until the destination is ready to accept the messages.

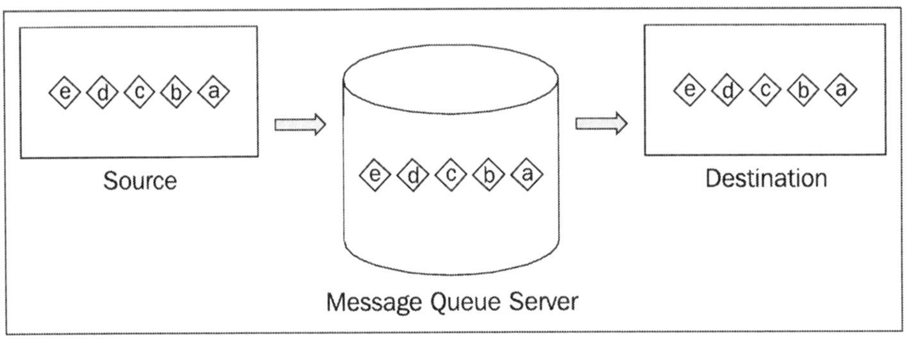

Message queuing has an important role in large-scaled distributed systems and enables asynchronous communication. Let's have a quick overview on the difference between synchronous and asynchronous systems.

In ordinary synchronous systems, tasks are processed one at a time. A task is not processed until the task in-process is finished. This is the simplest way to get the job done.

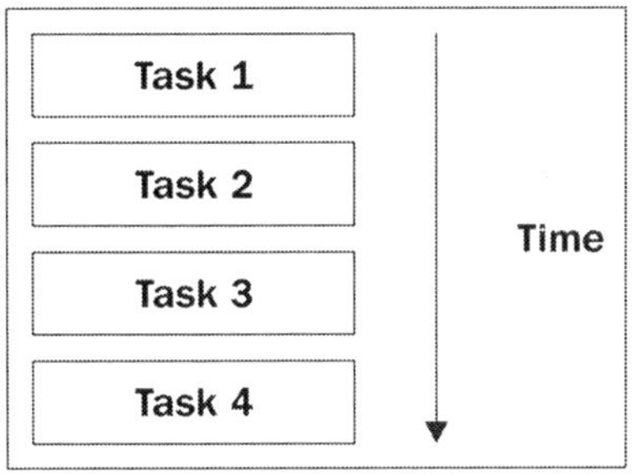

Synchronous system

We could also implement this system with threads. In this case threads process each task in parallel.

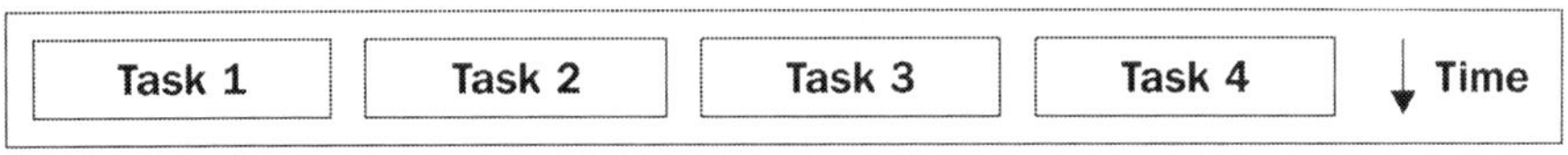

Threaded synchronous system

In the threading model, threads are managed by the operating system itself on a single processor or multiple processors/cores.

Asynchronous Input/Output (**AIO**) allows a program to continue its execution while processing input/output requests. AIO is mandatory in real-time applications. By using AIO, we could map several tasks to a single thread.

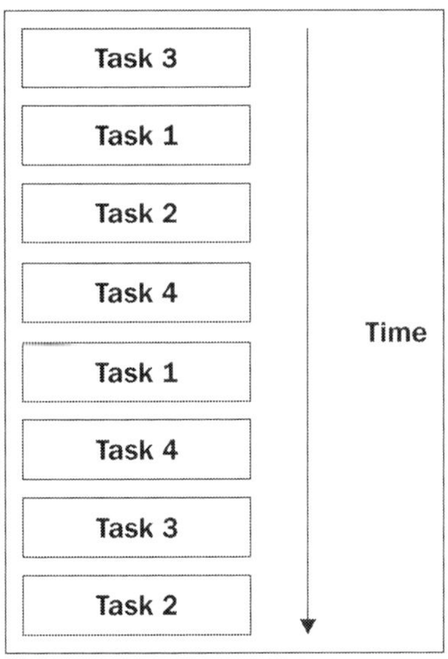

Asynchronous system

The traditional way of programming is to start a process and wait for it to complete. The downside of this approach is that it blocks the execution of the program while there is a task in progress. However, AIO has a different approach. In AIO, a task that does not depend on the process can still continue. We will cover AIO and how to use it with ZeroMQ in depth in *Chapter 2, Introduction to Sockets*.

You may wonder why you would use message queue instead of handling all processes with a single-threaded queue approach or multi-threaded queue approach. Let's consider a scenario where you have a web application similar to Google Images in which you let users type some URLs. Once they submit the form, your application fetches all the images from the given URLs. However:

- If you use a single-threaded queue, your application would not be able to process all the given URLs if there are too many users
- If you use a multi-threaded queue approach, your application would be vulnerable to a **distributed denial of service** attack (**DDoS**)
- You would lose all the given URLs in case of a hardware failure

In this scenario, you know that you need to add the given URLs into a queue and process them. So, you would need a message queuing system.

Introduction to ZeroMQ

Until now we have covered what a message queue is, which brings us to the purpose of this book, that is, **ZeroMQ**.

The community identifies ZeroMQ as "sockets on steroids". The formal definition of ZeroMQ is it is a messaging library that helps developers to design distributed and concurrent applications.

The first thing we need to know about ZeroMQ is that it is not a traditional message queuing system, such as ActiveMQ, WebSphereMQ, or RabbitMQ. ZeroMQ is different. It gives us the tools to build our own message queuing system. It is a library.

It runs on different architectures from ARM to Itanium, and has support for more than 20 programming languages.

Simplicity

ZeroMQ is simple. We can do some asynchronous I/O operations and ZeroMQ could queue the message in an I/O thread. ZeroMQ I/O threads are asynchronous when handling network traffic, so it can do the rest of the job for us. If you have worked on sockets before, you will know that it is quite painful to work on. However, ZeroMQ makes it easy to work on sockets.

Performance

ZeroMQ is fast. The website Second Life managed to get 13.4 microseconds end-to-end latencies and up to 4,100,000 messages per second. ZeroMQ can use multicast transport protocol, which is an efficient method to transmit data to multiple destinations.

The brokerless design

Unlike other traditional message queuing systems, ZeroMQ is brokerless. In traditional message queuing systems, there is a central message server (broker) in the middle of the network and every node is connected to this central node, and each node communicates with other nodes via the central broker. They do not directly communicate with each other.

However, ZeroMQ is brokerless. In a brokerless design, applications can directly communicate with each other without any broker in the middle. We will cover this topic in depth in *Chapter 2, Introduction to Sockets*.

> ZeroMQ does not store messages on disk. Please do not even think about it. However, it is possible to use a local swap file to store messages if you set `zmq.SWAP`.

Hello world

We can start writing some code after our introduction to message queuing and ZeroMQ and of course we will start with the famous "hello world" program.

Let's consider a scenario where we have a server and a client. The server replies `world` whenever it receives a `hello` message from the clients. The server runs on port `4040` and clients send messages to port `4040`.

The following is the server code, which sends the `world` message to clients:

```
#include <string.h>
#include <stdio.h>
#include <unistd.h>
#include "zmq.h"

int main (int argc, char const *argv[]) {

  void* context = zmq_ctx_new();
  void* respond = zmq_socket(context, ZMQ_REP);
  zmq_bind(respond, "tcp://*:4040");

  printf("Starting…\n");

  for(;;) {
    zmq_msg_t request;
    zmq_msg_init(&request);
    zmq_msg_recv(&request, respond, 0);
    printf("Received: hello\n");
    zmq_msg_close(&request);
    sleep(1); // sleep one second

    zmq_msg_t reply;
    zmq_msg_init_size(&reply, strlen("world"));
    memcpy(zmq_msg_data(&reply), "world", 5);
    zmq_msg_send(&reply, respond, 0);
    zmq_msg_close(&reply);
  }
  zmq_close(respond);
  zmq_ctx_destroy(context);

  return 0;
}
```

Downloading the example code

You can download the example code files for all Packt books you have purchased from your account at `http://www.PacktPub.com`. If you purchased this book elsewhere, you can visit `http://www.PacktPub.com/support` and register to have the files e-mailed directly to you.

The following is the client code that sends the `hello` message to the server:

```
#include <string.h>
#include <stdio.h>
#include <unistd.h>
#include "zmq.h"

int main (int argc, char const *argv[]) {

  void* context = zmq_ctx_new();

  printf("Client Starting....\n");

  void* request = zmq_socket(context, ZMQ_REQ);
  zmq_connect(request, "tcp://localhost:4040");

  int count = 0;

  for(;;) {
    zmq_msg_t req;
    zmq_msg_init_size(&req, strlen("hello"));
    memcpy(zmq_msg_data(&req), "hello", 5);
    printf("Sending: hello - %d\n", count);
    zmq_msg_send(&req, request, 0);
    zmq_msg_close(&req);

    zmq_msg_t reply;
    zmq_msg_init(&reply);
    zmq_msg_recv(&reply, request, 0);
    printf("Received: hello - %d\n", count);
    zmq_msg_close(&reply);
    count++;
  }
  // We never get here though.
  zmq_close(request);
  zmq_ctx_destroy(context);

  return 0;
}
```

> Please note that the examples in this book are written for ZeroMQ 3.2. Bear in mind that some examples may not work properly when using ZeroMQ Version 2.2 or older. Methods that were deprecated in 2.*x* were removed in 3.*x*. Some methods have been deprecated from those versions.

We have our first basic request-reply architecture, as shown in the following diagram:

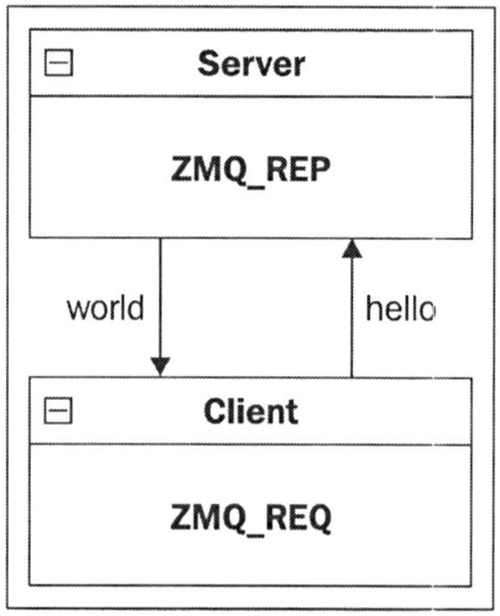

The request-reply pattern

Let's have a closer look at the code to understand how it works.

First we create a context and a socket. The `zmq_ctx_new()` method creates a new context. It is thread safe, so one context can be used from multiple threads.

`zmq_socket(2)` creates a new socket in the defined context. ZeroMQ sockets are not thread safe, so it should be used only by the thread where it was created. Traditional sockets are synchronous whereas ZeroMQ sockets have a queue on the client side and another on the server side for managing the request-reply pattern asynchronously. ZeroMQ automatically arranges setting up the connection, reconnecting, disconnecting, and content delivery. We will cover the difference between traditional sockets and ZeroMQ sockets in depth in *Chapter 3, Using Socket Topology*.

The server binds the `ZMQ_REP` socket to port `4040` and starts waiting for requests and replies back whenever it receives a message.

This basic "hello world" example introduces us to our first pattern, the request-reply pattern.

The request-reply pattern

We use the request-reply pattern to send messages from a client to one or multiple services and receive a reply for each message sent. This is most likely the easiest way to use ZeroMQ. The replies to the requests have to be strictly in order.

Reply

The following is the reply part of the request-reply pattern:

```
void* context = zmq_ctx_new();
void* respond = zmq_socket(context, ZMQ_REP);
zmq_bind(respond, "tcp://*:4040");
```

A server uses the `ZMQ_REP` socket to receive messages from and send replies to the clients. If the connection between a client and the server is lost then the replied message is thrown away without any notice. The incoming routing strategy of `ZMQ_REP` is fair-queue and the outgoing strategy is last-peer.

The fair-queue strategy

This book is all about queues. You may wonder what we mean when we refer to a fair-queue strategy. It is a scheduling algorithm and allocates the resources fairly by its definition.

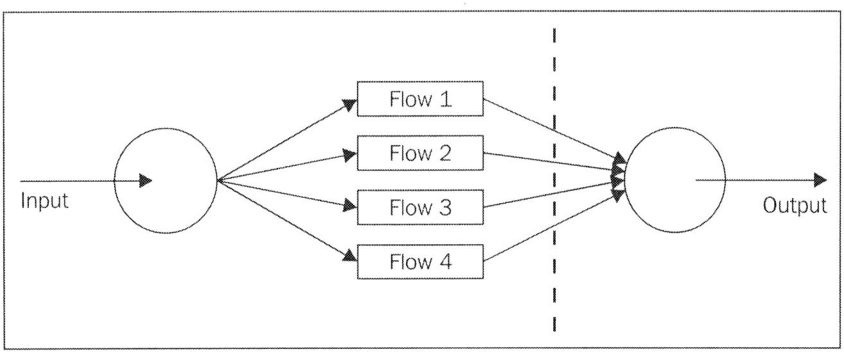

The fair-queue strategy

To understand how it works, let's say that the Flows in the preceding figure send 16, 2, 6, and 8 packets/second respectively, but the output can handle only 12 packets per second. In this case we could transmit 4 packets/second, but Flow 2 transmits only 2 packets/second. The rule of fair-queue is that there should not be any idle output unless all inputs are idle. Thus, we could allow Flow 2 to transmit its 2 packets/second and share the remaining 10 packets between the rest of the Flows.

This is the incoming routing strategy used by ZMQ_REP. The round-robin scheduling is the simplest way of implementing the fair-queue strategy, which is used by ZeroMQ as well.

Request

The following is the request part of the request-reply pattern:

```
void* context = zmq_ctx_new();
printf("Client Starting....\n");
void* request = zmq_socket(context, ZMQ_REQ);
zmq_connect(request, "tcp://localhost:4040");
```

A client uses ZMQ_REQ for sending messages to and receiving replies from a server. All messages are sent with the round-robin routing strategy. The incoming routing strategy is last-peer.

ZMQ_REQ does not throw away any messages. If there are no available services to send the message or if the all services are busy, all send operations—zmq_send(3)—are blocked until a service becomes available to send the message. ZMQ_REQ is compatible with the ZMQ_REP and ZMQ_ROUTER types. We will cover ZMQ_ROUTER in *Chapter 4, Advanced Patterns*.

Sending the message

This part combines the request and reply sections and shows how to request a message from somewhere and how to respond to them.

```
printf("Sending: hello - %d\n", count);
zmq_msg_send(&req, request, 0);
zmq_msg_close(&req);
```

The client sends the message to the server using zmq_msg_send(3). It queues the message and sends it to the socket.

```
int zmq_send_msg(zmq_msg_t *msg, void *socket, int flags)
```

zmq_msg_send takes three parameters, namely, message, socket, and flags.

- The message parameter is nullified during the request, so if you want to send the message to multiple sockets you need to copy it.
- A successful zmq_msg_send() request does not point out if the message has been sent over the network.

- The flags parameter is either `ZMQ_DONTWAIT` or `ZMQ_SNDMORE`. `ZMQ_DONTWAIT` indicates that the message should be sent asynchronously. `ZMQ_SNDMORE` indicates that the message is a multipart message and the rest of the parts of the message are on the way.

After sending the message, the client waits to receive a response. This is done by using `zmq_msg_recv(3)`.

```
zmq_msg_recv(&reply, request, 0);
printf("Received: hello - %d\n", count);
zmq_msg_close(&reply);
```

`zmq_msg_recv(3)` receives a part of the message from the socket, as specified in the socket parameter, and stores the reply in the message parameter.

```
int zmq_msg_recv (zmq_msg_t *msg, void *socket, int flags)
```

`zmq_msg_recv` takes three parameters, namely, message, socket, and flags.

- The previously received message (if any) is nullified
- The flags parameter could be `ZMQ_DONTWAIT`, which indicates that the operation should be done asynchronously

Handling strings in C

Every programming language has a different approach to handling strings. Erlang does not even have strings. In the C programming language, strings are null-terminated. Strings in C are basically character arrays where `\0` states the end of the string. String manipulation errors are common and the result of many security vulnerabilities.

According to Miller and others (1995), 65 percent of Unix failures are due to string manipulation errors such as null-terminated byte and buffer overflow; therefore, handling strings in C should be done carefully.

When you send a message with ZeroMQ, it is your responsibility to format it safely, so that other applications can read it. ZeroMQ only knows the size of the message. That's about it.

It is a common way to use different programming languages in an application. An application written in a programming language that does not add a null-byte at the end of strings and C application code needs to communicate properly otherwise you will get strange results.

You could send a message such as `world` as in our example with the null byte, as follows:

```
zmq_msg_init_data_(&request, "world", 6, NULL, NULL);
```

However, you would send the same message in Erlang as follows:

```
erlzmq:send(Request, <<"world">>)
```

Let's say our C client connects to a ZeroMQ service written in Erlang and we send the message `world` to this service. In this case Erlang will see it as `world`. If we send the message with the null byte, Erlang will see it as `[119,111,114,108,100,0]`. Instead of a string, we would get a list that contains some numbers! Well, those numbers are the ASCII-encoded characters. However, it is not interpreted as a string anymore.

You cannot rely on the fact that a message coming from a ZeroMQ service is safely terminated when you work in C.

Strings in ZeroMQ are fixed in length and are sent without the null byte. So, ZeroMQ strings are transmitted as some bytes (the string itself in this example) along with the length.

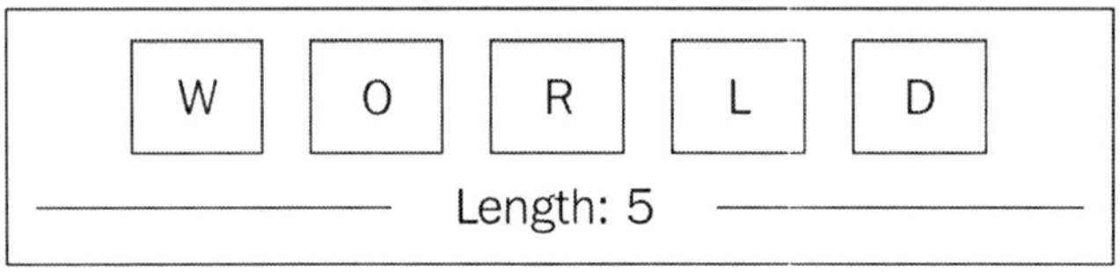

A ZeroMQ string

Checking the ZeroMQ version

It is quite useful to know which ZeroMQ version you are using. Knowing the exact version is helpful in some scenarios to avoid unwanted surprises. For example, there are some differences between ZeroMQ 2.*x* and ZeroMQ 3.*x*, such as deprecated methods; therefore, if you know the exact ZeroMQ version you have on your machine, you would avoid using deprecated methods.

```
#include <stdio.h>
#include "zmq.h"

int main (int argc, char const *argv[]) {
```

```
    int major, minor, patch;
    zmq_version(&major, &minor, &patch);
    printf("Installed ZeroMQ version: %d.%d.%d\n", major, minor,
           patch);

    return 0;
}
```

Summary

This chapter was an introduction to how message queuing works in general, then we had an introduction to ZeroMQ. We then looked at how ZeroMQ handles strings and introduced the request-reply pattern with a simple "hello world" application.

2

Introduction to Sockets

After having a look at the basic structure of ZeroMQ in previous chapter, in this chapter we will have a look at sockets with respect to the following points:

- The publish-subscribe pattern
- The pipeline pattern

The publish-subscribe pattern

First, let's introduce the second classic pattern, that is, the publish-subscribe pattern, which is a one-way distribution pattern where the server sends messages to a set of clients. It is a one-to-many model. The fundamental idea of this pattern is a publisher sends a message and connected subscribers receive the message, whereas disconnected subscribers just miss the message. A publisher is loosely coupled to the subscribers and does not care if any subscribers exist. It is similar to how TV channels or radio stations work. A TV channel always broadcasts TV shows and only the viewers who turn that channel on receive the broadcast. If you miss the time, you miss your favorite show (unless you have TiVo or something similar, but let's assume that our scenario happens in a world where recordings have not been invented). The advantage of the publish-subscribe pattern is that it provides a more dynamic network topology.

The publish-subscribe pattern can be summarized in the following four main points:

- **Publish**: An event is published by the publisher
- **Notify**: The subscriber is notified of a published event
- **Subscribe**: A new subscription is issued by a subscriber
- **Unsubscribe**: A subscriber removes its existing subscription

Let's take an example to make things clearer. Consider a scenario where we would like to set up a stock exchange program. There are brokers and they would like to know how certain stocks are doing in the market. Our publisher is the stock market and our subscribers are the brokers.

Instead of getting real numbers from stock markets, we will just generate some random numbers for stock values.

Before jumping into any code, first let's see what the publish-subscribe pattern looks like.

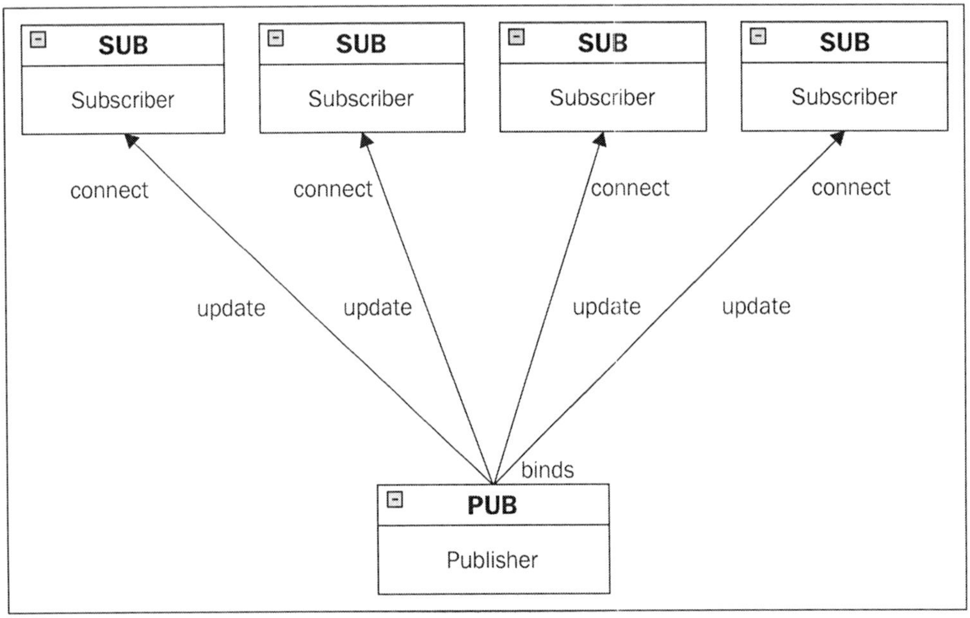

The publish-subscribe pattern

The following is the publisher code (server):

```
/*
 *  Stock Market Server
 *  Binds PUB socket to tcp://*:4040
 *  Publishes random stock values of random companies
 */

#include <string.h>
#include "zmq.h"
```

```
int main (int argc, char const *argv[]) {

  void* context = zmq_ctx_new();
  void* publisher = zmq_socket(context, ZMQ_PUB);
  printf("Starting server...\n");

  int conn = zmq_bind(publisher, "tcp://*:4040");

  const char* companies[2] = {"Company1", "Company2"};
  int count = 0;
  for(;;) {
    int price = count % 2;
    int which_company = count % 2;
    int index = strlen(companies[0]);
    char update[12];
    snprintf(update, sizeof update, "%s",
             companies[which_company]);

    zmq_msg_t message;
    zmq_msg_init_size(&message, index);
    memcpy(zmq_msg_data(&message), update, index);
    zmq_msg_send(&message, publisher, 0);
    zmq_msg_close(&message);
    count++;
  }

  zmq_close(publisher);
  zmq_ctx_destroy(context);

  return 0;
}
```

And the following is the subscriber code (client):

```
/*
 *  Stock Market Client
 *  Connects SUB socket to tcp://localhost:4040
 *  Collects stock exchange values
 */

#include <stdlib.h>
#include <string.h>
#include "zmq.h"
```

```
int main (int argc, char const *argv[]) {

  void* context = zmq_ctx_new();
  void* subscriber = zmq_socket(context, ZMQ_SUB);


  printf("Collecting stock information from the server.\n");

  int conn = zmq_connect(subscriber, "tcp://localhost:4040");
  conn = zmq_setsockopt(subscriber, ZMQ_SUBSCRIBE, 0, 0);

  int i;
  for(i = 0; i < 10; i++) {
    zmq_msg_t reply;
    zmq_msg_init(&reply);
    zmq_msg_recv(&reply, subscriber, 0);

    int length = zmq_msg_size(&reply);
    char* value = malloc(length);
    memcpy(value, zmq_msg_data(&reply), length);
    zmq_msg_close(&reply);
    printf("%s\n", value);
    free(value);
  }
  zmq_close(subscriber);
  zmq_ctx_destroy(context);

  return 0;
}
```

Setting a subscription using `zmq_setsockopt(3)` and `subscribe` is mandatory whenever you use a `SUB` socket, otherwise you will not receive any messages. This is a very common error.

The subscriber can set numerous subscriptions, which the subscriber receives, to any messages if an update matches any of the subscriptions. It can unsubscribe from particular subscriptions as well. Subscriptions are fixed-length blobs.

A subscriber receives the message using `zmq_msg_recv(3)`. `zmq_msg_recv(3)` receives a message from a socket and stores the message. Previous messages, if any, are deallocated.

```
int zmq_msg_recv (zmq_msg_t *msg, void *socket, int flags);
```

The flag option can only be one value, which is `ZMQ_DONTWAIT`. If `ZMQ_DONTWAIT` is specified, then the operation is performed in the non-blocking mode. If the message is successfully received, it (`zmq_msg_recv(3)`) returns the size of the message in bytes; otherwise it returns `-1` and the error message flag.

The publish-subscribe pattern is asynchronous and sending a message to a `SUB` socket causes an error. You could call `zmq_msg_send(3)` to send messages whenever you want but you should never call `zmq_msg_recv(3)` on a `PUB` socket.

The following is the sample output of the client code:

```
Company2 570
Company2 878
Company2 981
Company2 783
Company1 855
Company1 524
Company2 639
Company1 984
Company1 158
Company2 145
```

The publisher will always send messages even if there is no subscriber. You could try it and see for yourself. You would see that the publisher sends something like the following:

```
Sending... Company2 36
Sending... Company2 215
Sending... Company2 712
Sending... Company2 924
Sending... Company2 721
Sending... Company1 668
Sending... Company2 83
Sending... Company2 209
Sending... Company1 450
Sending... Company1 940
Sending... Company1 57
Sending... Company2 3
Sending... Company1 100
Sending... Company2 947
```

Let's say we want to receive the results of `Company1` or the company name that we pass as an argument. In that case, we could change our client code to the following:

```
//
//  Stock Market Client
//  Connects SUB socket to tcp://localhost:4040
//  Collects stock exchange values
//

#include <stdlib.h>
#include <string.h>
#include "zmq.h"


int main (int argc, char const *argv[]) {

  void* context = zmq_ctx_new();
  void* subscriber = zmq_socket(context, ZMQ_SUB);

  const char* filter;

  if(argc > 1) {
    filter = argv[1];
  } else {
    filter = "Company1";
  }
  printf("Collecting stock information from the server.\n");

  int conn = zmq_connect(subscriber, "tcp://localhost:4040");
  conn = zmq_setsockopt(subscriber, ZMQ_SUBSCRIBE, filter,
                        strlen(filter));

  int i = 0;
  for(i = 0; i < 10; i++) {
    zmq_msg_t reply;
    zmq_msg_init(&reply);
    zmq_msg_recv(&reply, subscriber, 0);

    int length = zmq_msg_size(&reply);
    char* value = malloc(length + 1);
    memcpy(value, zmq_msg_data(&reply), length);
    zmq_msg_close(&reply);
```

```
        printf("%s\n", value);
        free(value);

    }
    zmq_close(subscriber);
    zmq_ctx_destroy(context);

    return 0;
}
```

The output would be something like the following:

```
Company1 575
Company1 504
Company1 513
Company1 584
Company1 444
Company1 1010
Company1 524
Company1 963
Company1 929
Company1 718
```

Filtering out messages

Our basic stock exchange application sends messages to subscribers. It seems everything has gone as expected, right? Unfortunately, no.

ZeroMQ matches the subscriber strings with prefix matching, which means ZeroMQ will return `Company1`, `Company10`, and `Company101` even if you look for `Company1` alone.

Let's change our publisher code to the following:

```
//
//  Stock Market Server
//  Binds PUB socket to tcp://*:4040
//  Publishes random stock values of random companies
//

#include <string.h>
#include "zmq.h"
```

```
int main (int argc, char const *argv[]) {

  void* context = zmq_ctx_new();
  void* publisher = zmq_socket(context, ZMQ_PUB);

  int conn = zmq_bind(publisher, "tcp://*:4040");

  const char* companies[3] = {"Company1", "Company10",
                              "Company101"};

  int count = 0;
  for(;;) {
    int price = count % 17;
    int which_company = count % 3;
    int index = strlen(companies[which_company]);
    char update[64];
    snprintf(update, sizeof update, "%s",
             companies[which_company]);

    zmq_msg_t message;
    zmq_msg_init_size(&message, index);
    memcpy(zmq_msg_data(&message), update, index);
    zmq_msg_send(&message, publisher, 0);
    zmq_msg_close(&message);
    count++;

  }

  zmq_close(publisher);
  zmq_ctx_destroy(context);

  return 0;
}
```

And now let's change our subscriber code to the following:

```
//
//  Stock Market Client
//  Connects SUB socket to tcp://localhost:4040
//  Collects stock exchange values
//

#include <stdlib.h>
#include <string.h>
```

```
#include "zmq.h"


int main (int argc, char const *argv[]) {

  void* context = zmq_ctx_new();
  void* subscriber = zmq_socket(context, ZMQ_SUB);

  const char* filter;

  if(argc > 1) {
    filter = argv[1];
  } else {
    filter = "Company1";
  }
  printf("Collecting stock information from the server.\n");

  int conn = zmq_connect(subscriber, "tcp://localhost:4040");
  conn = zmq_setsockopt(subscriber, ZMQ_SUBSCRIBE, filter,
                        strlen(filter));

  int i = 0;
  for(i = 0; i < 10; i++) {
    zmq_msg_t reply;
    zmq_msg_init(&reply);
    zmq_msg_recv(&reply, subscriber, 0);

    int length = zmq_msg_size(&reply);
    char* value = malloc(length + 1);
    memcpy(value, zmq_msg_data(&reply), length);
    zmq_msg_close(&reply);
    printf("%s\n", value);
    free(value);

  }
  zmq_close(subscriber);
  zmq_ctx_destroy(context);

  return 0;
}
```

In this case, the output will be something similar to the following:

```
Collecting stock information from the server.
Company101 950
Company10 707
Company101 55
Company101 343
Company10 111
Company1 651
Company10 287
Company101 8
Company1 889
Company101 536
```

Our subscriber code explicitly says that we want to see the results of `Company1`. However, the publisher sends us the results of `Company10` and `Company101` as well. This is certainly not what we want. We need to solve this small issue.

We may want to do some dirty hacking to get what we want but using a delimiter is a much simpler solution for it.

We need to make some changes both in the publisher and the subscriber code and we will filter the company names using a delimiter.

The following is our updated publisher code that fixes the previous problem. Have a look at the highlighted line to see how we can use a delimiter to send the message to the subscribers:

```
//
//  Stock Market Server
//  Binds PUB socket to tcp://*:4040
//  Publishes random stock values of random companies
//

#include <stdlib.h>
#include <string.h>
#include "zmq.h"


int main (int argc, char const *argv[]) {

  void* context = zmq_ctx_new();
  void* publisher = zmq_socket(context, ZMQ_PUB);
```

```
  int conn = zmq_bind(publisher, "tcp://*:4040");
  conn = zmq_bind(publisher, "ipc://stock.ipc");

  const char* companies[3] = {"Company1", "Company10",
                               "Company101"};

  for(;;) {
    int price = count % 17;
    int which_company = count % 3;
    int index = strlen(companies[which_company]);
    char update[64];
    sprintf(update, "%s| %d", companies[which_company], price);
    zmq_msg_t message;
    zmq_msg_init_size(&message, index);
    memcpy(zmq_msg_data(&message), update, index);
    zmq_msg_send(&message, publisher, 0);
    zmq_msg_close(&message);
    count++;
  }

  zmq_close(publisher);
  zmq_ctx_destroy(context);

  return 0;
}
```

And the following is our updated subscriber code to filter results using the delimiter we use in our publisher code:

```
//
//  Stock Market Client
//  Connects SUB socket to tcp://localhost:4040
//  Collects stock exchange values
//

#include <stdlib.h>
#include <string.h>
#include "zmq.h"


int main (int argc, char const *argv[]) {
```

```
    void* context = zmq_ctx_new();
    void* subscriber = zmq_socket(context, ZMQ_SUB);

    const char* filter;

    filter = "Company1|";
    printf("Collecting stock information from the server.\n");

    int conn = zmq_connect(subscriber, "tcp://localhost:4040");
    conn = zmq_setsockopt(subscriber, ZMQ_SUBSCRIBE, filter,
                          strlen(filter));

    int i = 0;
    for(i = 0; i < 10; i++) {
      zmq_msg_t reply;
      zmq_msg_init(&reply);
      zmq_msg_recv(&reply, subscriber, 0);

      int length = zmq_msg_size(&reply);
      char* value = malloc(length + 1);
      memcpy(value, zmq_msg_data(&reply), length);
      zmq_msg_close(&reply);
      printf("%s\n", value);
      free(value);
    }
    zmq_close(subscriber);
    zmq_ctx_destroy(context);

    return 0;
  }
```

Now we can see the results as expected after the changes we have made in our publisher and subscriber code.

The socket options

Since we use the publish-subscribe pattern, the option name we use is `ZMQ_SUBSCRIBE`.

```
int conn = zmq_connect(subscriber, "tcp://localhost:4040");
conn = zmq_setsockopt(subscriber, ZMQ_SUBSCRIBE, option_value,
                      strlen(option_value));
```

The socket options are set with the `zmq_setsockopt(3)` function. It takes four parameters:

- Socket
- Option name
- Option value
- Length of the option

This can be made clear by the following line of code:

```
int zmq_setsockopt (void *socket, int option_name, const void *option_
value, size_t option_len);
```

Subscription

`ZMQ_SUBSCRIBE` establishes a new message on the `ZMQ_SUB` socket. If the `option_value` argument is not empty, we are subscribed to all messages that start with `option_value`. You could attach multiple filters to a one `ZMQ_SUB` socket.

Unsubscription

`ZMQ_UNSUBSCRIBE` removes a message on the `ZMQ_SUB` socket. It removes only one message even if there are multiple filters.

An important thing we need to note about the publisher-subscriber sockets is that we do not know when the subscriber starts to receive messages. In this case, it is a good idea to start the subscriber and then to start the publisher. This is because the subscriber always misses the first message as connecting to the publisher takes time and the publisher may already be sending a message.

However, we will talk about how to synchronize the publisher and the subscribers so we do not have to send any messages unless the subscribers are really connected.

Notes on the publisher-subscriber pattern

The key points to be noted about the publisher-subscriber pattern are as follows:

- Messages are queued up on the publisher's side if you are using TCP and the subscriber is too slow to receive messages. We will show you how to protect the application against this.

- A subscriber could connect to multiple publishers. Data will be transmitted via the fair-queue strategy.
- The publisher sends all the messages to all subscribers and filtering is done on the subscriber's side as we have seen from our stock exchange program that we provided earlier.
- We will return to the publisher-subscriber pattern in *Chapter 4, Advanced Patterns*, to discuss how to deal with the slow subscribers.

The pipeline pattern

Let's continue with the pipeline pattern. The pipeline pattern transmits data between nodes ordered in the pipeline. Data is transmitted continuously and each step of the pipeline is connected to one or more nodes. A round-robin strategy is used to transmit data between nodes. It is somewhat similar to the request-reply pattern.

The divide and conquer strategy

It is like there is no escape from a divide and conquer strategy when you do programming. Remember when you enrolled in your algorithms class and your annoying professor introduced divide and conquer using merge sort and a week later half of the class dropped the unit? We remember as well. It is a small world and here is divide and conquer, again.

Let's do something in parallel with ZeroMQ. Consider a scenario where we have a producer that generates some random numbers. We also have workers, which find the square root of those numbers with Newton's method. Then we have a collector that collects the results from the workers.

The following is our server code:

```
#include <stdio.h>
#include <stdlib.h>
#include <unistd.h>
#include <string.h>
#include <sys/time.h>
#include <time.h>
#include "zmq.h"

int main (int argc, char const *argv[]) {

  void* context = zmq_ctx_new();
```

```
    // This is the socket that we send messages.
    void* socket = zmq_socket(context, ZMQ_PUSH);
    zmq_bind(socket, "tcp://*:4040");

    // This is the socket that we send batch message.
    void* connector = zmq_socket(context, ZMQ_PUSH);
    zmq_connect(connector, "tcp://localhost:5050");

    printf("Please press enter when workers are ready...");
    getchar();
    printf("Sending tasks to workers...\n");

    // The first message. It's also the signal start of batch.
    int length = strlen("-1");
    zmq_msg_t message;
    zmq_msg_init_size(&message, length);
    memcpy(zmq_msg_data(&message), "-1", length);
    zmq_msg_send(&message, connector, 0);
    zmq_msg_close(&message);

    // Generate some random numbers.
    srandom((unsigned) time(NULL));

    // Send the tasks.
    int count;
    int msec = 0;
    for(count = 0; count < 100; count++) {
      int load = (int) ((double) (100) * random () / RAND_MAX);
      msec += load;
      char string[10];
      sprintf(string, "%d", load);
    }
    printf("Total: %d msec\n", msec);
    sleep(1);

    zmq_close(connector);
    zmq_close(socket);
    zmq_ctx_destroy(context);

    return 0;
}
```

The following is the worker code where we do some square root calculations using Newton's method:

```
#include <stdlib.h>
#include <string.h>
#include <unistd.h>
#include "zmq.h"

double square(double x) {
  return x * x;
}

double average(double x, double y) {
  return (x + y) / 2.0;
}

double good_enough(double guess, double x) {
  return abs(square(guess) - x) < 0.000001;
}

double improve(double guess, double x) {
  return average(guess, x / guess);
}

double sqrt_inner(double guess, double x) {
  if(good_enough(guess, x))
    return guess;
  else
    return sqrt_inner(improve(guess, x), x);
}

double newton_sqrt(double x) {
  return sqrt_inner(1.0, x);
}

int main (int argc, char const *argv[]) {

  void* context = zmq_ctx_new();

  // Let's initialize a socket to receive messages.
  void* receiver = zmq_socket(context, ZMQ_PULL);
  zmq_connect(receiver, "tcp://localhost:4040");
```

```
    // Let's initialize a socket to send the messages.
    void* sender = zmq_socket(context, ZMQ_PUSH);
    zmq_connect(sender, "tcp://localhost:5050");

    for(;;) {
      zmq_msg_t reply;
      zmq_msg_init(&reply);
      zmq_msg_recv(&reply, receiver, 0);

      int length = zmq_msg_size(&reply);
      char* msg = malloc(length + 1);
      memcpy(msg, zmq_msg_data(&reply), length);
      zmq_msg_close(&reply);

      fflush(stdout);
      double val = atof(msg);
      printf("%.1f: %.1f\n", val, newton_sqrt(val));

      sleep(1);
      free(msg);

      zmq_msg_t message;
      char* ssend = "T";
      int t_length = strlen(ssend);
      zmq_msg_init_size(&message, t_length);
      memcpy(zmq_msg_data(&message), ssend, t_length);
      zmq_msg_send(&message, receiver, 0);
      zmq_msg_close(&message);

    }
    zmq_close(receiver);
    zmq_close(sender);
    zmq_ctx_destroy(context);

    return 0;
  }
```

The following is the receiver code:

```
  #include <stdlib.h>
  #include <string.h>
  #include "zmq.h"
```

```
int main (int argc, char const *argv[]) {

  void* context = zmq_ctx_new();
  void* receiver = zmq_socket(context, ZMQ_PULL);
  zmq_bind(receiver, "tcp://*:5050");

  // We receive the first message and discard it since it's the
  // signal start of batch which is -1.
  zmq_msg_t reply;
  zmq_msg_init(&reply);
  zmq_msg_recv(&reply, receiver, 0);

  int length = zmq_msg_size(&reply);
  char* msg = malloc(length + 1);
  memcpy(msg, zmq_msg_data(&reply), length);
  zmq_msg_close(&reply);
  free(msg);

  int count;
  for(count = 0; count < 100; count++) {
    zmq_msg_t reply;
    zmq_msg_init(&reply);
    zmq_msg_recv(&reply, receiver, 0);

    int length = zmq_msg_size(&reply);
    char* value = malloc(length + 1);
    memcpy(value, zmq_msg_data(&reply), length);
    zmq_msg_close(&reply);
    free(value);
    if(count / 10 == 0)
      printf("10 Tasks have been processed.");
    fflush(stdout);
  }

  zmq_close(receiver);
  zmq_ctx_destroy(context);

  return 0;
}
```

The following diagram represents the code we have written so far:

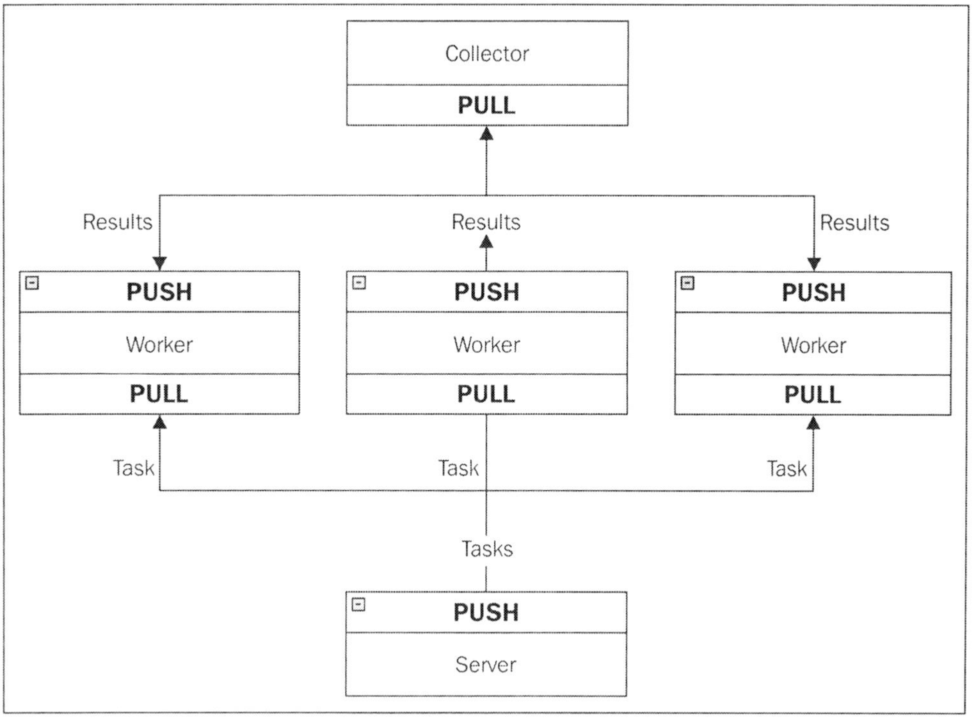

What we have done so far:

- The start of the batch needs to be synchronized when the workers are up and running. As we have said earlier, the connection process takes some time. If we do not do that, then the first worker will take messages while other workers are being connected. In order to prevent this, we need to synchronize the start of the batch to run in parallel.
- The collector's **PULL** socket fetches the results using the fair-queue scheduling that we described in *Chapter 1, Getting Started*.
- The server's **PUSH** socket sends the tasks to the workers evenly.
- The workers are connected downstream to the collector and upstream to the server. You could add more workers explicitly.

We have mentioned that workers connect to downstream to the collector and upstream to the server. Now, let's examine what this means more closely.

Let's have a look at the following code snippet from our worker code:

```
// Let's initialize a socket to receive messages.
  void* receiver = zmq_socket(context, ZMQ_PULL);
  zmq_connect(receiver, "tcp://localhost:4040");
```

The ZMQ_PULL socket

When we want to retrieve data from upstream to nodes, we use ZMQ_PULL. The ZMQ_PULL type sockets are used to receive messages from upstream nodes in the pipeline. As we have said earlier, this process is done with fair-queue scheduling.

> zmq_send(3) cannot be used in place of ZMQ_PULL.

The ZMQ_PUSH socket

When we want to communicate downstream with nodes, we use ZMQ_PUSH. The ZMQ_PUSH type sockets are used to send messages to downstream nodes in the pipeline.

ZMQ_PUSH never discards messages. If a high watermark is reached for downstream nodes, or if there are no downstream nodes available, all messages sent with zmq_send(3) are blocked until there is an available downstream node to receive messages.

Getting ZeroMQ context

You must have realized by now that all examples we have done so far start with zmq_ctx_new(3). ZeroMQ applications always start with creating a context. All sockets are put inside a single process using the context, which acts as in-process sockets since they are the fastest way to connect threads in a single process. ZeroMQ context is thread safe and you may share it with multiple threads.

> If ZeroMQ context cannot be created, it returns NULL.

Even though it is possible to create multiple contexts, which would be considered as separate ZeroMQ applications, it is a better idea to create one ZeroMQ context rather than multiple ones.

Destroying ZeroMQ context

At the end of your application code, you need to destroy the context you have created by calling `zmq_ctx_destroy(3)`. Once `zmq_ctx_destroy(3)` is called, all processes are returned with an error code (`ETERM`) and `zmq_ctx_destroy(3)` blocks the calls until opened sockets within the context are closed by `zmq_close(3)`.

Cleaning up

When you code in a programming language such as Python or Java, you do not need to worry about memory management since these languages have built-in mechanisms to clean up the memory.

For example, Python uses reference counting when there are cycles in a reference chain of objects, which never gets freed and memory automatically gets cleaned up for you when you implement a ZeroMQ application in Python. So, you do not need to explicitly close a connection when coding a ZeroMQ application since it will be closed as soon as the reference count of the object becomes zero. However, we should note that this would not work on Jython, PyPy, or IronPython. Anyway, this is enough information on Python. Let's return to our main concern.

When you code in C, memory management is your responsibility. Otherwise you will have an unstable application, which will have memory leak problems.

You need to take care of sockets, messages, and contexts in ZeroMQ. There are a couple of things you need to consider to finish an application successfully:

- As we have said earlier, you need to close the application by destroying the context using `zmq_ctx_destroy(3)`. However, if there are some sockets open, `zmq_ctx_destroy(3)` will just wait there forever. Therefore, you need to close the sockets and then call `zmq_ctx_destroy(3)` to destroy the context.
- `zmq_ctx_destroy(3)` will wait forever if there are connections or messages on the queue to be sent.
- Whenever you have finished processing a message, you need to close it immediately by calling `zmq_msg_close(3)`, otherwise your application may have memory leaks. You could think about it this way: when you leave your house, you close the door. You do not leave it open. Similar idea.
- Do not open and close a lot of sockets. If you do so, it pretty much means you are doing something wrong and you will want to redesign your application from scratch.

You may wonder what happens if your application is a multithreaded application. Well, in that case, things get really complicated.

Detecting memory leaks

Memory management needs to be done carefully when an application is coded in C or C++, since it is the developer's responsibility to manage the memory. For this purpose, we are going to use a Linux-only tool called **Valgrind**. It can be used to detect memory leaks or generate profiling data, among many other useful sanity checks on the running code.

Firstly, the following section is a small tutorial on Valgrind where we will discuss using Valgrind with ZeroMQ.

Introduction to Valgrind

You could compile your program with the `-g` parameter to include debugging information. In that case, error messages will include exact line numbers. Using `-O1` can result in inaccurate messages, and using `-O2` or `-O3` definitely results in inaccurate messages.

Consider the following example:

```
#include <stdio.h>
#include <stdlib.h>

int main(int argc, char const *argv[]) {

  char* a = malloc(4);
  int b;
  printf("b = %d\n", b);

  return 0;
}
```

Let's compile with `gcc -g -o test test.c`. Now, it is time to run Valgrind to check for memory leaks. Let's run the following command:

```
valgrind --leak-check=full --show-reachable=yes test
```

Once we run the previous command, Valgrind will check for memory errors using the `memcheck` tool. You could specify it using `tool=memcheck`, but this would be pointless in this case since `memcheck` is the default tool. The output would be something similar to the following:

```
==98190== Conditional jump or move depends on uninitialised value(s)
==98190==    at 0x2D923: __vfprintf
==98190==    by 0x4AC5A: vfprintf_l
==98190==    by 0x952BE: printf
==98190==    by 0x1F5E: main (test.c:8)

==98190== 4 bytes in 1 blocks are definitely lost in loss record 1 of 5
==98190==    at 0xF656: malloc (vg_replace_malloc.c:195)
==98190==    by 0x1F46: main (test.c:6)

==98190== LEAK SUMMARY:
==98190==    definitely lost: 4 bytes in 1 blocks
==98190==    indirectly lost: 0 bytes in 0 blocks
==98190==      possibly lost: 0 bytes in 0 blocks
```

Let's now describe the preceding output:

- We could ignore `98190` since it is the process ID
- `Conditional jump or move depends on uninitialised value(s)` means we have uninitialized variables in our code
- `definitely lost` means there is a memory leak and we need to fix it
- `indirectly lost` means the blocks that point to another block are lost
- `possibly lost` means there is a memory leak, unless you really know what you are doing

By default, Valgrind uses `$PREFIX/lib/valgrind/default.supp`. However, we need to create our own suppression file to use with ZeroMQ, which would be something like the following:

```
{
   <socketcall_sendto>
   Memcheck:Param
```

```
      socketcall.sendto(msg)
      fun:send
      ...
   }
   {
      <socketcall_sendto>
      Memcheck:Param
      socketcall.send(msg)
      fun:send
      ...
   }
```

Then you could run Valgrind with arguments similar to the following:

```
valgrind --leak-check=full --show-reachable=yes --suppressions=zeromq.
supp server
```

Here we can say that running code under Valgrind has severe impacts on the performance and can introduce timeouts in some cases.

Summary

In this chapter, we had an introduction to sockets and we introduced two new patterns, namely the publish-subscribe and pipeline pattern. We also discussed how to solve a certain problem using the publish-subscribe pattern with the help of a simple example. Then we had a small discussion on how to detect memory leaks in our ZeroMQ applications by using Valgrind.

3
Using Socket Topology

We have looked at basic patterns such as pipeline, publish-subscribe, request-reply, detecting memory leaks, and also how to deal with the borderline of publish and subscribe. In this chapter, we are going to dig deep into sockets. First, let's start with a description of sockets.

What a socket is

Sockets are the de-facto standard API for performing network programming. The socket API is very similar to the file I/O in many aspects since we perform `open`, `write`, `read`, and `close` operations. They let us interact with other nodes in the network. There are several different kinds of sockets and they have different properties. The two most common families are:

- `PF_UNIX` for Unix inter-process communication
- `PF_INET` for Internet communication

However, we are going to look at DARPA Internet addresses (Internet sockets) since the most common ZeroMQ sockets are Internet sockets.

Types of Internet sockets

There are different types of Internet sockets. You may have seen `SOCK_DGRAM`, `SOCK_STREAM`, and `SOCK_RAW` before. The following are the brief definitions of the most popular ones:

- **Stream sockets** (`SOCK_STREAM`): These types of sockets use Transmission Control Protocol (TCP) or Stream Control Transmission Protocol (SCTP). It ensures that sent data is sequenced and unduplicated and it is a reliable socket. A sample usage would be `socket(PF_INET, SOCK_STREAM, 0);`.

- **Datagram sockets** (`SOCK_DGRAM`): These types of sockets use User Datagram Protocol (UDP). These sockets are known as connectionless sockets. It should be noted that they are unreliable and the data may arrive out of sequence. There may be duplication as well. A sample usage would be `socket(PF_INET, SOCK_DGRAM, 0);`.
- **Raw sockets** (`SOCK_RAW`): These types of socket neither use TCP nor UDP. It directly communicates with the IP layer. This may be useful to build a new protocol and is a lower-level approach.

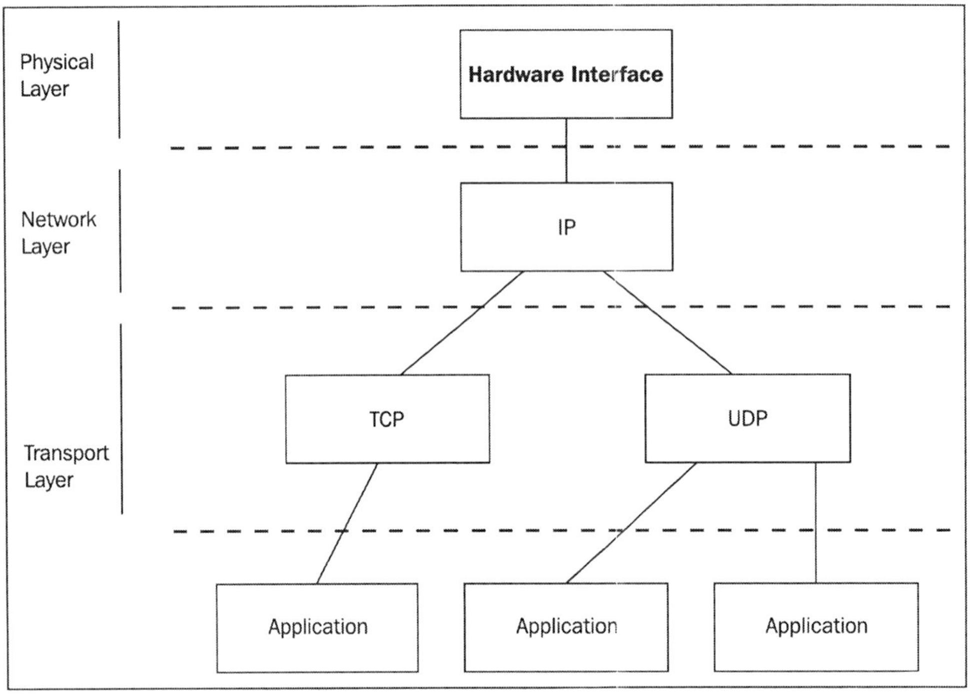

The network layer

Our main concern is the `SOCK_STREAM` type of sockets. These types of sockets are unduplicated and have a best-effort policy to guarantee in-order delivery of datagrams to have a full reconstruction of the stream on the receiving side. What we can do with these sockets is very simple:

- Create a connection with `connect`
- Use `read` and `write` or `send` and `recv` to transmit data respectively
- Close the connection with `close`

Transmission Control Protocol (TCP)

TCP prevents loss of data, duplication, and damage. It ensures that the sent message isn't out of order. These characteristics make it a reliable protocol, unlike UDP.

> Please refer to the *RFC 793* guide for more information about TCP.

TCP synchronizes using a three-way handshake. Synchronization between two nodes begins when a TCP segment is sent with the `SYN (0x02)` flag set.

TCP sockets are one-to-one. A source node transmits messages to the destination node.

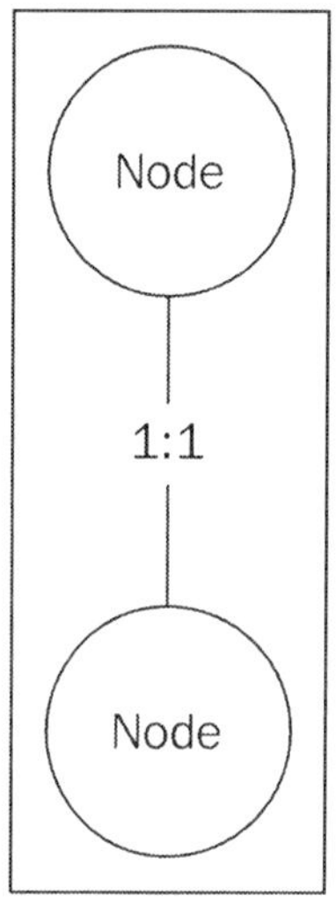

TCP sockets are one-to-one

The three-way handshake protocol

A three-way handshake allows TCP to prevent out-of-order message delivery and duplicated transmission.

1. Transmission begins when the `SYN (0x02)` flag is set in a packet and is sent to the destination.

2. The destination receives the packet with the `SYN (0x02)` flag and sends an acknowledgement to the source by setting the `ACK (0x10)` flag in the reply. This stage is called SYN-ACK.
3. The source receives the SYN-ACK message and sends an `ACK` segment, which notes that each packet in the TCP messages contains a sequence number to maintain the order.
4. Once the transmission is done, the `FIN (0x01)` flag is set in the next reply packet.

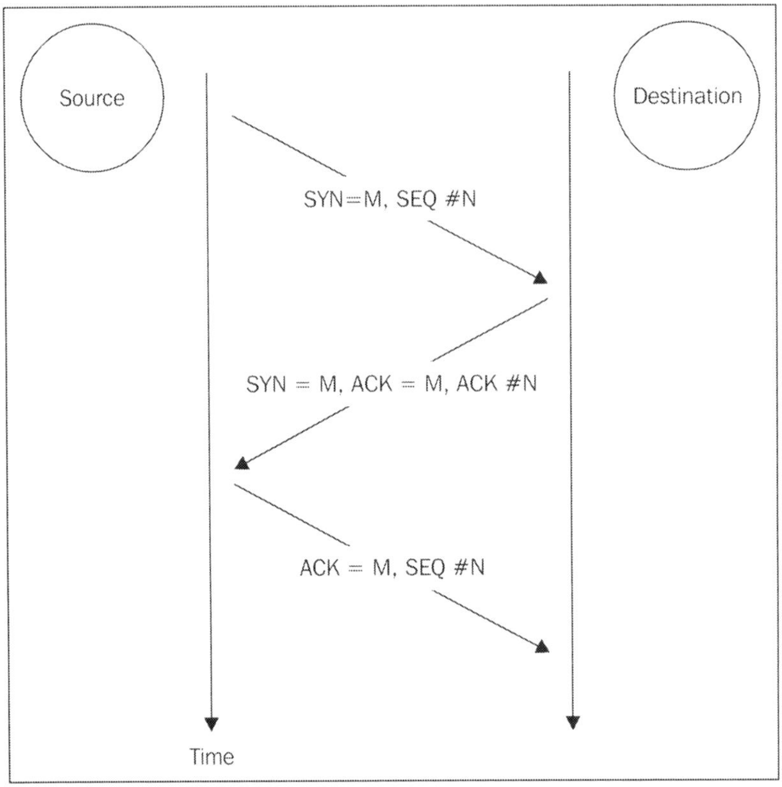

The three-way handshake protocol

The sequence numbers allow TCP to control the flow. TCP uses a sliding-window technique. If an `ACK` segment is not received in a particular period of time, then it retransmits the missing part of data. However, choosing an appropriate timeout is not an easy task. Therefore, in order to solve this issue, TCP uses adaptive retransmission.

TCP header

The following diagram shows the header format of TCP:

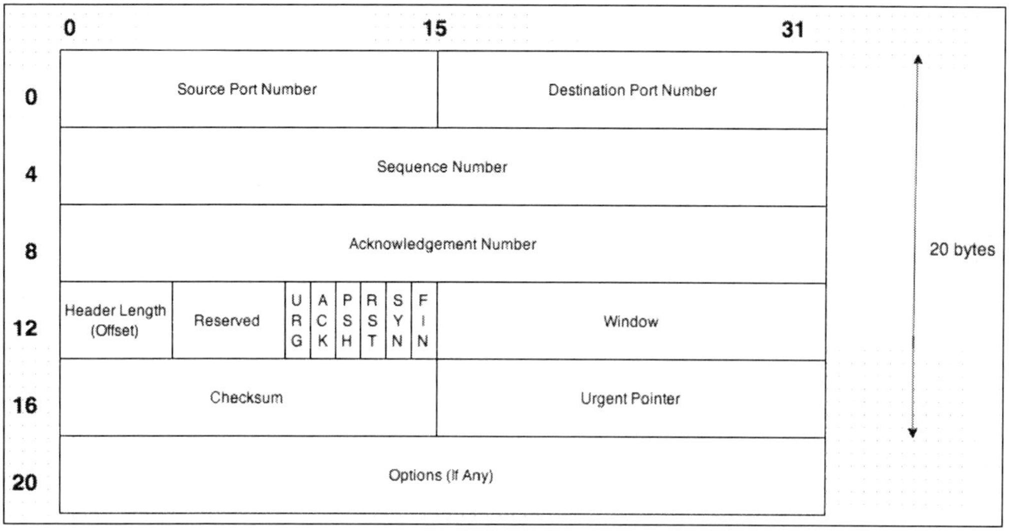

TCP header

The source port number and destination port number set the port numbers. The sequence number starts at the fourth byte in the message transmitted from the source node to the destination node, as shown in the figure demonstrating the three-way handshake protocol. The acknowledgement number is the next sequence number that the source node expects.

TCP flags

The following is the list of flags that are commonly used in ZeroMQ:

- `URG (0x20)`: Urgent
- `ACK (0x10)`: Acknowledgement
- `PSH (0x08)`: Push
- `RST (0x04)`: Reset
- `SYN (0x02)`: Synchronize
- `FIN (0x01)`: Finish

Properties of TCP

Some of the properties of TCP that make it popular among the users are as follows:

- **Reliability**: This is the most important characteristic of TCP. As we have said earlier, TCP guarantees unduplicated transmission and data delivery.
- **Full-duplex**: TCP allows full-duplex operations, hence two nodes can send messages to each other simultaneously.
- **Flow control**: TCP uses the sliding-window technique to implement flow control. Sequence numbers attached to the transmission travels with the `ACK (0x10)` flag set. This indicates the number of bytes they can receive without resulting in a buffer overflow.
- **Multiplexing**: This feature allows a single node to handle multiple processes simultaneously.

ZeroMQ sockets

ZeroMQ sockets have four different methods, just like normal sockets, as we have said at the beginning of this chapter. It should be noted that ZeroMQ sockets are always void pointers and are asynchronous.

Differences between TCP sockets and ZeroMQ sockets

The following is the difference between TCP sockets and ZeroMQ sockets:

- ZeroMQ sockets are asynchronous.
- They may implement particular patterns.
- TCP sockets are one-to-one whereas ZeroMQ sockets are many-to-many. However, you could implement one-to-many, one-to-one, many-to-one, or many-to-many with ZeroMQ, depending on your needs and socket type.
- ZeroMQ sockets transmit messages whereas TCP transmit bytes. As we said in *Chapter 1, Getting Started*, a message is a fixed-length binary object.
- I/O is done in the background in ZeroMQ sockets. Even if your application is too busy to handle messages, they are put in queues.

- Unlike TCP, ZeroMQ sockets do not care whether the destination exists or not.
- ZeroMQ sockets may transmit data to multiple nodes and receive data from multiple nodes.

We cannot start the source node and then the destination node when we work with TCP sockets. This is because the source node would immediately try to connect to the destination and if there was no destination present to receive the message, then we would have a problem. However, in ZeroMQ, the message can be enqueued and sent later if there is no destination to receive the message. Another difference is that we tell ZeroMQ to send and receive messages instead of bytes.

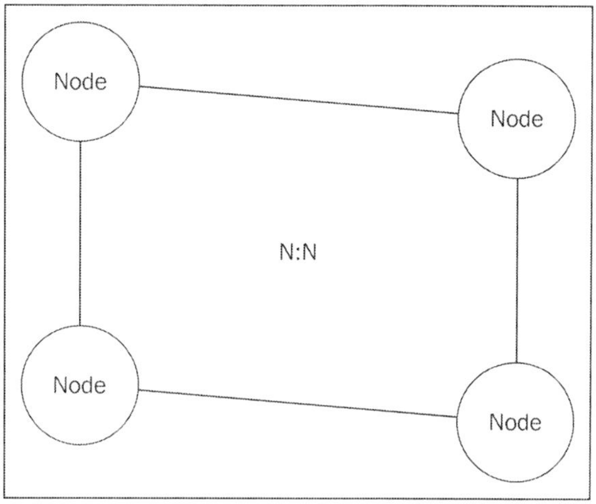

ZeroMQ sockets are many-to-many

Routing schemes

We will focus mainly on the unicast scheme, but there are other routing schemes worth mentioning:

- **Unicast**: This is the major message transmission scheme among the other routing schemes. It has a one-to-one relationship where the source transmits a message to only one destination.
- **Multicast**: This is a one-to-many approach where one source transmits messages to many destinations, which are subscribed to the source. It does not guarantee the delivery of messages to the destinations, just like UDP.

- **Broadcast**: This is a one-to-all approach where the source transmits messages to every single destination. However, not all protocols support broadcasting (for example, X.25).
- **Geocast**: This is a one-to-many approach where one source transmits messages to multiple destinations based on their geographic locations.

ZeroMQ supports unicast transports such as IPC, TCP, and INPROC and multicast transports such as **Pragmatic General Multicast (PGM)** and **Encapsulated Pragmatic General Multicast (EPGM)**. EPGM datagrams are encapsulated inside UDP datagrams.

ZeroMQ is bundled with OpenPGM. This is the advantage of using ZeroMQ over AMQP since AMQP has no multicast support. Multicast allows us to send data to all clients only once, hence the bandwidth usage is kept steady. However, this should not be taken as a comparison between ZeroMQ and AMQP. This would not be a healthy comparison. As we said in *Chapter 1, Getting Started*, ZeroMQ is a library.

> PGM and EPGM can only be used with `ZMQ_SUB` and `ZMQ_PUB`.

Unicast

Unicast is a transmission scheme that transmits messages to one destination. ZeroMQ supports TCP, INPROC, and IPC.

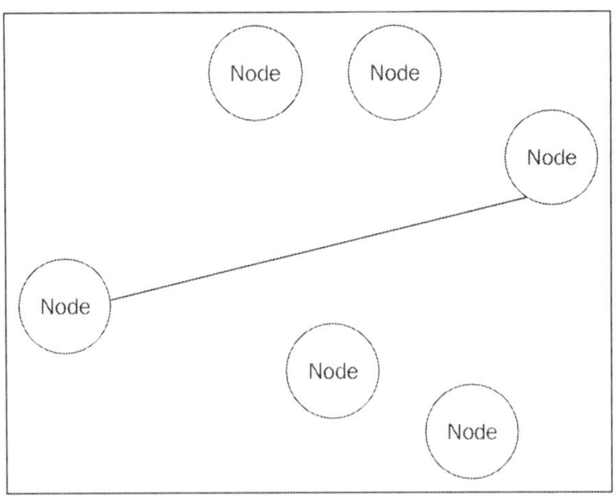

Unicast scheme

Inter-process communication (IPC) is the transmission of data among processes or threads.

TCP transport is another option you could use. However, TCP transportation of ZeroMQ does not care whether a destination node exists or not. This is similar to IPC.

Inter-thread transport (INPROC) has a restriction; it is mandatory to bind before creating a connection.

Setting I/O threads and limiting the number of sockets

ZeroMQ does I/O in the background. You could change the number of threads to work with `zmq_ctx_set` by setting it before creating a socket. ZeroMQ creates one thread by default.

```
void* context = zmq_ctx_new();
zmq_ctx_set(context, ZMQ_IO_THREADS, 8);
```

Limiting the number of sockets would be a good idea since ZeroMQ will continue to create sockets as long as your operating system can handle it and this will make your application vulnerable to the denial of service attacks.

```
void* context = zmq_ctx_new();
zmq_ctx_set(context, ZMQ_MAX_SOCKETS, 512);
```

Working with multiple sockets

In *Chapter 1, Getting Started*, and *Chapter 2, Introduction to Sockets*, we worked on programs with a single socket. Working on one socket is easy, but working on multiple sockets is somewhat tricky. To work with multiple sockets, we use `zmq_poll(3)`, which is an event loop that allows an application to multiplex I/O with multiple sockets.

```
/*

  Polling with ZeroMQ

*/
```

```
#include <string.h>
#include "zmq.h"

int main (int argc, char const *argv[]) {

  void* ctx = zmq_ctx_new();
  void* pull = zmq_socket(ctx, ZMQ_PULL);
  zmq_connect(pull, "tcp://localhost:4040");

  void* subscriber = zmq_socket(ctx, ZMQ_SUB);

  char* company_name = "Company1";
  int length = strlen(company_name) + 1;

  zmq_connect(subscriber, "tcp://localhost:5050");
  zmq_setsockopt(subscriber, ZMQ_SUBSCRIBE, company_name, length);

  printf("starting...\n");

  zmq_pollitem_t polls[2];

  polls[0].socket = pull;
  polls[0].fd = 0;
  polls[0].events = ZMQ_POLLIN;
  polls[0].revents = 0;

  polls[1].socket = subscriber;
  polls[1].fd = 0;
  polls[1].events = ZMQ_POLLIN;
  polls[1].revents = 0;

  for(;;) {
    zmq_msg_t msg;
    int res =  zmq_poll(polls, 2, -1);

    if(polls[0].revents > 0) {
      zmq_msg_init(&msg);
      zmq_msg_recv(&msg, pull, 0);
      zmq_msg_close(&msg);
    }

    if(polls[1].revents > 0) {
      zmq_msg_init(&msg);
```

```
        zmq_msg_recv(&msg, subscriber, 0);
        zmq_msg_close(&msg);
      }
    }

    zmq_close(pull);
    zmq_close(subscriber);
    zmq_ctx_destroy(ctx);

    return 0;
  }
```

You could initialize polling options by putting them into an array as well. For example, the following code connects to three different sockets and pulls (`ZMQ_PULL`) the results from those sockets. Those sockets push data using `ZMQ_PUSH`.

```
/*

 Pull from multiple sockets with zmq_poll.

*/

#include "zmq.h"

int main (int argc, char const *argv[]) {

  void* context = zmq_ctx_new();

  void* pull1 = zmq_socket(context, ZMQ_PULL);
  zmq_bind(pull1, "tcp://*:5050");

  void* pull2 = zmq_socket(context, ZMQ_PULL);
  zmq_bind(pull2, "tcp://*:4040");

  void* pull3 = zmq_socket(context, ZMQ_PULL);
  zmq_bind(pull3, "tcp://*:6060");

  printf("Starting...\n");

  zmq_pollitem_t polls[] = {
    {pull1, 0, ZMQ_POLLIN, 0},
    {pull2, 0, ZMQ_POLLIN, 0},
```

```
    {pull3, 0, ZMQ_POLLIN, 0}
  };

  int length = sizeof(polls) / sizeof(zmq_pollitem_t);

  for(;;) {
    zmq_msg_t msg;
    zmq_poll(polls, length, -1);

    if(polls[0].revents & ZMQ_POLLIN) {
      zmq_msg_init(&msg);
      zmq_msg_recv(&msg, pull1, 0);
      zmq_msg_close(&msg);
    }
    if(polls[1].revents > 0) {
      zmq_msg_init(&msg);
      zmq_msg_recv(&msg, pull2, 0);
      zmq_msg_close(&msg);
    }
    if(polls[2].revents > 0) {
      zmq_msg_init(&msg);
      zmq_msg_recv(&msg, pull3, 0);
      zmq_msg_close(&msg);
    }
  }

  zmq_close(pull1);
  zmq_close(pull2);
  zmq_close(pull3);
  zmq_ctx_destroy(context);

  return 0;
}
```

Never access the `zmq_msg_t` variables directly.

Working with multi-part messages

We always define messages using zmq_msg. When we want to send multi-part messages, again, we need to use zmq_msg. For example, if the data package is divided into 10 parts, you need to create 10 zmq_msg sockets. The client either receives all the message parts or nothing at all. In order to send multi-part messages, the ZMQ_SNDMORE flag must be set during the zmq_send call.

```
int zmq_send(void* socket, void* buf, size_t len, int flags);
```

The following is the request-reply example that we used in *Chapter 1, Getting Started,* but this time we will send the message in multiple parts. First, let's have a look at the server code:

```
/*

  Request - Reply

  Send "world" in multiple-parts.

  server.c

*/

#include <string.h>
#include <stdio.h>
#include <unistd.h>
#include "zmq.h"

int main (int argc, char const *argv[]) {

  void* context = zmq_ctx_new();
  void* respond = zmq_socket(context, ZMQ_REP);
  zmq_bind(respond, "tcp://*:4040");

  printf("Starting...\n");

  for(;;) {

    zmq_msg_t request;
    zmq_msg_init(&request);
    zmq_msg_recv(&request, respond, 0);
```

```
        printf("Received: hello\n");

        zmq_msg_close(&request);

        sleep(1);

        zmq_msg_t msg1, msg2, msg3, msg4, msg5;
        zmq_msg_init_size(&msg1, 2);
        zmq_msg_init_size(&msg2, 2);
        zmq_msg_init_size(&msg3, 2);
        zmq_msg_init_size(&msg4, 2);
        zmq_msg_init_size(&msg5, 2);

        memcpy(zmq_msg_data(&msg1), "w", 2);
        zmq_msg_send(&msg1, respond, ZMQ_SNDMORE);


        memcpy(zmq_msg_data(&msg2), "o", 2);
        zmq_msg_send(&msg2, respond, ZMQ_SNDMORE);

        memcpy(zmq_msg_data(&msg3), "r", 2);
        zmq_msg_send(&msg3, respond, ZMQ_SNDMORE);

        memcpy(zmq_msg_data(&msg4), "l", 2);
        zmq_msg_send(&msg4, respond, ZMQ_SNDMORE);

        memcpy(zmq_msg_data(&msg5), "d", 2);
        zmq_msg_send(&msg5, respond, 0);


        zmq_msg_close(&msg1);
        zmq_msg_close(&msg2);
        zmq_msg_close(&msg3);
        zmq_msg_close(&msg4);
        zmq_msg_close(&msg5);
    }

    zmq_close(respond);
    zmq_ctx_destroy(context);

    return 0;
}
```

And the following is the client code:

```
/*

  Request - Reply

  Receive "world" in multi-parts.

  client.c

*/


#include <string.h>
#include <stdio.h>
#include <stdint.h>
#include <unistd.h>
#include "zmq.h"

int main (int argc, char const *argv[]) {

  void* context = zmq_ctx_new();

  printf("Client Starting....\n");

  void* request = zmq_socket(context, ZMQ_REQ);
  zmq_connect(request, "tcp://localhost:4040");


  for(;;) {

    zmq_msg_t reply;
    zmq_msg_init(&reply);
    zmq_msg_recv(&reply, request, 0);
    printf("Received: %s\n", (char *) zmq_msg_data(&reply));
    zmq_msg_close(&reply);

    uint64_t more_part;
    size_t more_size = sizeof(more_part);
    zmq_getsockopt(request, ZMQ_RCVMORE, &more_part, &more_size);
```

```
        if (!more_part)
          break;

    }

    zmq_close(request);
    zmq_ctx_destroy(context);

    return 0;
  }
```

We should note that the last part of the message flag that you send should always be set to `0`. If you set it to `ZMQ_SNDMORE`, then the client will not receive any messages.

```
zmq_msg_t msg1, msg2, msg3, msg4, msg5;
zmq_msg_init_size(&msg1, 2);
zmq_msg_init_size(&msg2, 2);
zmq_msg_init_size(&msg3, 2);
zmq_msg_init_size(&msg4, 2);
zmq_msg_init_size(&msg5, 2);


zmq_msg_send(&msg1, respond, ZMQ_SNDMORE);
zmq_msg_send(&msg2, respond, ZMQ_SNDMORE);
zmq_msg_send(&msg3, respond, ZMQ_SNDMORE);
zmq_msg_send(&msg4, respond, ZMQ_SNDMORE);
zmq_msg_send(&msg5, respond, ZMQ_SNDMORE);
```

The previous code snippet will not work. You could change the relevant parts of server code with it and experience it yourself.

How to handle interruptions

You need to close the applications properly when you interrupt your application with signals such as `SIGTERM` or `SIGINT`. `SIGTERM` is one of the POSIX signals that sends a signal to a process to end it. Signals are asynchronous and are vulnerable to race conditions.

- `SIGTERM`: When you execute the `kill` command with its defaults in a Unix system, you are basically calling a `SIGTERM` signal to the denoted process. These signals should be handled properly so your application can release the resources or close database connections and flush the messages properly.

- `SIGINT`: This signal can be caught by the application just like `SIGTERM` as well. The user usually calls it with *Ctrl* + *C*.
- `SIGKILL`: This is the signal that is called with a `kill -9` command. However, this signal cannot be caught by an application, so there is nothing much we could do about it.

`SIGINT` could be caught in languages that support exception handling by throwing the proper exception such as `KeyboardInterrupt` in Python. However, things are a little bit different in C.

Let's recall our "hello world" request-reply program to demonstrate the handling of the `SIGTERM` (*Ctrl* + *C*) interrupt.

```
/*

  Request - Reply

  Handling interrupts.

  server.c

*/

#include <string.h>
#include <stdio.h>
#include "czmq.h"

int main (int argc, char const *argv[]) {

  zctx_t* context = zctx_new();
  void* socket = zsocket_new(context, ZMQ_REP);
  zsocket_bind(socket, "tcp://*:5050");

  printf("Starting server...\n");

  for(;;) {

    char* msg = zstr_recv(socket);

    if(!msg) {
      break;
    }
```

```
    printf("Received: %s\n", msg);
    zstr_send(socket, "world");
    free(msg);
  }

  zsocket_destroy(context, socket);
  zctx_destroy(&context);

  return 0;
}
```

And the following is the client code:

```
/*

  Request - Reply

  Handling interrupts.

  client.c

*/


#include "czmq.h"

int main (int argc, char const *argv[]) {

  zctx_t* context = zctx_new();
  void* request = zsocket_new(context, ZMQ_REQ);

  printf("Starting client...\n");

  zsocket_connect(request, "tcp://localhost:5050");

  for(;;) {
    zstr_send(request, "hello");
    char* reply = zstr_recv(request);
    if(!reply) {
      break;
    }
```

```
        printf("Received: %s\n", reply);
        free(reply);
        sleep(1);
    }

    zsocket_destroy(context, request);
    zctx_destroy(&context);

    return 0;

}
```

However, we are coding in C and it already has signal-handling functions defined in `signal.h`. We could do things closer to the traditional C style by including a signal handler, as follows:

```
/*

    Request - Reply

    Send "world" in multi-parts.

    server.c

*/

#include "czmq.h"
#include <signal.h>

int main (int argc, char const *argv[]) {

    zctx_t* context = zctx_new();
    void* socket = zsocket_new(context, ZMQ_REP);
    zsocket_bind(socket, "tcp://*:5050");
    signal(SIGINT, exit);

    printf("Starting server...\n");

    for(;;) {

        char* msg = zstr_recv(socket);
```

```
        printf("Received: %s\n", msg);
        zstr_send(socket, "world");
        free(msg);
    }

    zsocket_destroy(context, socket);
    zctx_destroy(&context);

    return 0;
}
```

Did you notice something different? Yes, we included the `czmq.h` header. The code has completely changed.

Introduction to CZMQ

CZMQ is a high-level C library for ZeroMQ. Its main purpose is to minimize the differences between ZeroMQ v2.*x* and v3.*x* and also to enable doing more with less code. If you recall from *Chapter 1, Getting Started*, we said that the examples are written for ZeroMQ v3.2. Therefore, they may not work with ZeroMQ v2.2 or older.

It includes list and hash structures and lets developers work with strings and multi-part messages easier. It also automatically closes opened sockets when a given context is terminated.

Linking CZMQ with your ZeroMQ application is simple:

```
gcc -lczmq -o program program.c
```

zctx

This is a wrapper for the ZeroMQ context. Its main features are as follows:

- Setting up signal handling, thus blocking calls such as `zmq_poll` and `zmq_recv()` and returning when `SIGINT` (*Ctrl* + *C*) or `SIGTERM` (`kill`) is called.
- It automatically closes the opened sockets before terminating the context.

Sample usage:

```
zctx_t* context = zctx_new();
```

zstr_send

This is called to send and receive C strings. It is used to send basic strings. The important thing we need to note is that `zstr` sends strings without a trailing null byte, `\0`. However, it appends a null byte after receiving the string.

Sample usage:

```
zstr_send(socket, "world");
```

zloop

This provides an event-driven reactor pattern by handling events using `zmq_pollitem_t`. The event-driven programming paradigm is useful since it does not require a new thread for each request. Threading may result in poor performance because of context switching.

Here is a simple example. The server receives a "hello" message from a client and sends back the message.

First, let's see the server code:

```
/*
  Polling example

*/

#include "czmq.h"
#include <signal.h>

int do_something(zloop_t* loop, zmq_pollitem_t* item, void* socket) {
  char* s = zstr_recv(socket);
  printf("%s\n", s);
  return 0;
}

int main (int argc, char const *argv[]) {

  zctx_t* context = zctx_new();
  void* socket = zsocket_new(context, ZMQ_PULL);
  zsocket_bind(socket, "tcp://*:5050");
  signal(SIGINT, exit);
```

```
  printf("Starting server...\n");

  zloop_t* loop = zloop_new();

  zloop_set_verbose(loop, 1);
  zloop_timer(loop, 10000, 1, do_send, NULL);

  zmq_pollitem_t poll = {socket, 0, ZMQ_POLLIN};
  zloop_poller(loop, &poll, do_something, socket);

  zloop_start(loop);
  zloop_destroy(&loop);

  zsocket_destroy(context, socket);
  zctx_destroy(&context);

  return 0;
}
```

And the following is the client code:

```
#include "czmq.h"
#include <signal.h>

int main (int argc, char const *argv[]) {

  zctx_t* context = zctx_new();
  void* request = zsocket_new(context, ZMQ_PUSH);
  signal(SIGINT, exit);
  printf("Starting client...\n");

  zsocket_connect(request, "tcp://localhost:5050");
  int i = 0;

  for(;;) {
    zstr_send(request, "hello");
    printf("Pushing Hello\n");
}

  zsocket_destroy(context, request);
  zctx_destroy(&context);

  return 0;

}
```

zmsg

This is used to work with multi-part messages. Its sample usage is as follows:

```
/*

  Send "hello" in multi-parts using zmsg.

  server.c

*/

#include "czmq.h"

int main (int argc, char const *argv[]) {

  zctx_t* context = zctx_new();
  void* socket = zsocket_new(context, ZMQ_REP);
  zsocket_bind(socket, "tcp://*:5050");

  printf("Starting server...\n");

  for(;;) {

    zmsg_t* msg = zmsg_new();

    zmsg_addmem(msg, "h", 1);
    zmsg_addmem(msg, "e", 1);
    zmsg_addmem(msg, "l", 1);
    zmsg_addmem(msg, "l", 1);
    zmsg_addmem(msg, "o", 1);

    zmsg_send(&msg, socket);

    zmsg_destroy(&msg);
  }

  zsocket_destroy(context, socket);
  zctx_destroy(&context);

  return 0;
}
```

zfile

This contains the following helper functions to work with files:

- `zfile_size`: This detects the file size.
- `zfile_mkdir`: This creates a directory if it does not exist. However, it goes only one level deep; therefore, it will create `/my_zmq_folder/` but will not create `/my_zmq_files/my_another_folder/`.
- `zfile_delete`: This deletes the file.
- `zfile_exists`: This detects if files already exist or not.

Here is some sample code for `zfile`. The scenario is that a client receives the "world" message from the server and writes it to a file.

The following server code is identical to what we have done in the interrupt signals example:

```
/*

  Request - Reply

  Working with files

  Helloserver.c

*/

#include "czmq.h"

int main (int argc, char const *argv[]) {

  zctx_t* context = zctx_new();
  void* socket = zsocket_new(context, ZMQ_REP);
  zsocket_bind(socket, "tcp://*:5050");

  printf("Starting server...\n");

  for(;;) {

    char* msg = zstr_recv(socket);
```

```
        if(!msg) {
          break;
        }

        printf("Received: %s\n", msg);
        zstr_send(socket, "world");
        free(msg);
      }

      zsocket_destroy(context, socket);
      zctx_destroy(&context);

      return 0;
    }
```

The client sends a "Hello" message to the server, receives the "world" message, and writes the "world" message to a file:

```
    /*

      Request - Reply

      Receive "world" in multi-parts.

      client.c

    */


    #include <stdio.h>
    #include "czmq.h"

    int main (int argc, char const *argv[]) {

      zctx_t* context = zctx_new();
      void* request = zsocket_new(context, ZMQ_REQ);

      printf("Starting client...\n");

      zsocket_connect(request, "tcp://localhost:5050");

      zstr_send(request, "Hello");
```

```
  zmsg_t* msg = zmsg_recv(request);


  FILE* file = fopen("server.txt", "w");
  zmsg_save(msg, file);
  fclose(file);

  zmsg_destroy(&msg);

  zsocket_destroy(context, request);
  zctx_destroy(&context);

  return 0;

}
```

zfile_mkdir

As we have said earlier, `zfile_mkdir` creates a directory if and only if it is one level deep. Therefore, it will create a `test` folder, but will not create a `test/test2` folder. Here it is in action:

```
/*

  Request - Reply

  Working with zfile_mkdir


*/


#include <stdio.h>
#include "czmq.h"

int main (int argc, char const *argv[]) {

  zctx_t* context = zctx_new();
  void* request = zsocket_new(context, ZMQ_REQ);

  printf("Starting client...\n");
```

```
    zsocket_connect(request, "tcp://localhost:5050");

    zstr_send(request, "Hello");


    zfile_mkdir("test"); // Will work

    zfile_mkdir("test2/test3"); // Will not work


    zmsg_destroy(&msg);
    zsocket_destroy(context, request);
    zctx_destroy(&context);

    return 0;

}
```

zhash

This is basically a hash table. It internally uses a Bernstein hash data structure to hash the strings. You could use `zhash` for caching; for example, the client sends a key to the server and receives the e-mail address of a user.

The following is the server code:

```
/*

    Request - Reply

    Using zhash.

    server.c

*/

#include "czmq.h"
#include <signal.h>

int main (int argc, char const *argv[]) {

    zctx_t* context = zctx_new();
```

```
    void* socket = zsocket_new(context, ZMQ_REP);
    zsocket_bind(socket, "tcp://*:5050");
    signal(SIGINT, exit);

    printf("Starting server...\n");

    zhash_t* map = zhash_new();
    zhash_insert(map, "user_id", "1234");
    zhash_insert(map, "user_email", "name@example.org");

    for(;;) {

      char* msg = zstr_recv(socket);

      char* email = zhash_lookup(map, msg);
      printf("Received: %s\n", msg);

      if(email) {
        zstr_send(socket, zhash_lookup(map, msg));
      } else {
        zstr_send(socket, "Not Found");
      }

      free(msg);
    }

    zsocket_destroy(context, socket);
    zctx_destroy(&context);

    return 0;
  }
```

And the client code is as follows:

```
  /*

    Request - Reply

    Using zhash.
```

```
  client.c

*/
#include "czmq.h"
#include <signal.h>

int main (int argc, char const *argv[]) {

  zctx_t* context = zctx_new();
  void* request = zsocket_new(context, ZMQ_REQ);

  printf("Starting client...\n");

  zsocket_connect(request, "tcp://localhost:5050");

  for(;;) {

    /* Send 'user_email' key to the server and receive the e-mail
       address of a user.
    */
    zstr_send(request, "user_email");
    char* reply = zstr_recv(request);

    if(!reply) {
      break;
    }

    printf("Received: %s\n", reply);

    free(reply);
    sleep(1);
  }

  zsocket_destroy(context, request);
  zctx_destroy(&context);

  return 0;

}
```

zlist

This is a single-list interface. For example, you could use `zlist` to store buffer results. It is reasonably fast.

The following is a client example. The server example provided for `zhash` can be used with this:

```
#include "czmq.h"
#include <signal.h>

int main (int argc, char const *argv[]) {

  zctx_t* context = zctx_new();
  void* request = zsocket_new(context, ZMQ_REQ);

  printf("Starting client...\n");

  zsocket_connect(request, "tcp://localhost:5050");

  zlist_t* list = zlist_new();


  zstr_send(request, "user_email");
  char* reply = zstr_recv(request);

  zlist_push(list, reply);

  zstr_send(request, "user_id");
  reply = zstr_recv(request);

  zlist_push(list, reply);

  int length = zlist_size(list);
  for(i = 0; i < length; i++) {
    char* s = (char *) zlist_pop(list);
    printf("%s ", s);
  }
  printf("\n");

  sleep(1);

  free(reply);
```

```
    zsocket_destroy(context, request);
    zctx_destroy(&context);

    return 0;

}
```

zclock

This has sleep and timing functions. You would use this mostly for testing and debugging threads or polls.

zthread

This is used to create detached threads. Such threads create their own ZeroMQ context, `zctx_t`.

> There is also the `libzfl` library, which aims to build industrial-scale ZeroMQ applications that you could use. The `libzfl` library relies on CZMQ, so you need to have CZMQ before using `libzfl`.

Summary

In this chapter, we discussed how ZeroMQ sockets differ from traditional TCP sockets. Later on, we discussed limiting the number of sockets and setting I/O threads. We also discussed how to work with multiple sockets and how we could send multi-part messages. In the last section of this chapter, we learned about CZMQ, a high-level C binding library for ZeroMQ, which would minimize the difference between ZeroMQ v2.*x* and ZeroMQ v3.*x*.

4
Advanced Patterns

In this chapter we will look at multithreading applications with ZeroMQ and have a brief look at more advanced patterns. This chapter is a journey into ZeroMQ's advanced features.

Extending the request-reply pattern

In *Chapter 1, Getting Started*, we worked on a simple request-reply example. When you have one server to handle multiple client responses, everything is fine. However, in a realistic situation, there would be more than one server. How do we solve this, then? We could set up a broker to handle it. However, ZeroMQ is brokerless. Perhaps we could use `zmq_poll`, but in this case we will not be able to use the request-reply pattern since it only gives us synchronous transmission.

It looks like we are stuck, but there must be some way to solve this. If you remember from *Chapter 1, Getting Started*, we said that ZeroMQ gives us tools to code our own message-queuing service. We are going to implement a broker to solve this small issue.

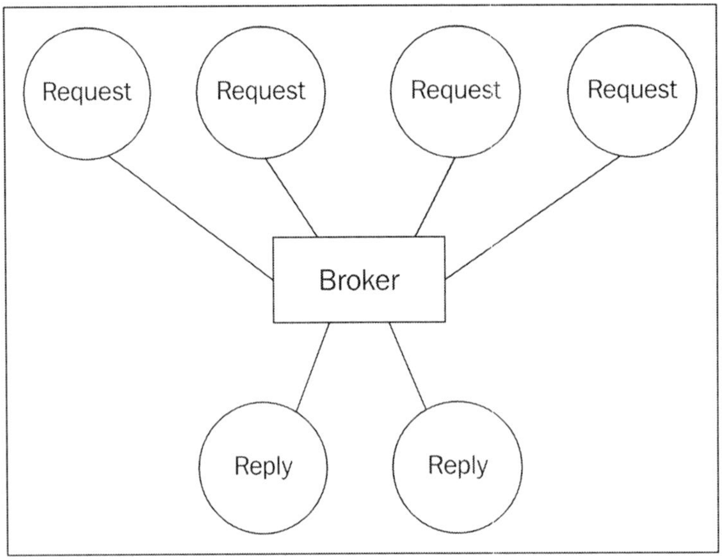

Extended request-reply pattern

When there is a broker, servers and clients do not interact directly with each other. They communicate only through a broker.

This is a brief introduction to more advanced topics that we will cover later in this chapter.

Writing multithreaded applications with ZeroMQ

ZeroMQ threads are native threads, which means they are OS threads instead of green threads. The difference between native threads and green threads is that the latter run on user space whereas the former run on kernel space. The main disadvantage of green threads is that the OS has no clue what is going on since these kinds of threads do not rely on the operating system.

There are a few things you need to keep in mind when programming multithreaded applications with ZeroMQ:

- As mentioned in the previous chapters, ZeroMQ sockets are not thread safe. Therefore, you should not share sockets between threads.
- Only ZeroMQ context is thread safe.
- Do not access the same messages from different threads.

The following is a sample multithreaded "Hello world" server using `pthread`.

```
#include "czmq.h"
#include <pthread.h>
#include <stdio.h>
#include <stdlib.h>


void* worker(void* ctx) {
  zctx_t* context = ctx;
  void* receiver = zsocket_new(context, ZMQ_REP);
  zsocket_connect(receiver, "inproc://workers");

  for(;;) {
    char* str = zstr_recv(receiver);
    printf("Received: %s\n", str);
    free(str);

    zclock_sleep(10);
    zstr_send(receiver, "world");
  }

  zsocket_destroy(context, receiver);
  zmq_close(receiver);
  return NULL;
}


int main (int argc, char const *argv[]) {

  zctx_t* context = zctx_new();
  void* clients = zsocket_new(context, ZMQ_ROUTER);
  zsocket_bind(clients, "tcp://*:4040");
```

```
    void* workers = zsocket_new(context, ZMQ_DEALER);
    zsocket_bind(workers, "inproc://workers");

    int i;
    for(i = 0; i < 5; i++) {
      pthread_t thread;
      pthread_create(&thread, NULL, worker, context);
    }
    zclock_sleep(10);

    zctx_destroy(&context);
    return 0;
  }
```

Wrapping publisher-subscriber messages

If you recall from *Chapter 2, Introduction to Sockets,* we said that messages are prefix matched when using the publisher-subscriber pattern and we have showed a work-around to get what the subscriber really wants. This time, we visit the PUB-SUB sockets by enveloping messages with separate keys.

The following is the server code:

```
/*

  PUB - SUB wrap messages.
  server.c

*/

#include "czmq.h"


int main (int argc, char const *argv[]) {

  zctx_t* context = zctx_new();
  void* pub = zsocket_new(context, ZMQ_PUB);
  zsocket_bind(pub, "tcp://*:4040");

  printf("starting server\n");
```

```
    for(;;) {
      zstr_sendm(pub, "Company1");
      zstr_send(pub, "Company Message to be ignored.");
      zstr_sendm(pub, "Company10");
      zstr_send(pub, "Company message to receive.");
      zclock_sleep(10);
    }

    zsocket_destroy(context, pub);
    zctx_destroy(&context);

    return 0;
  }
```

And the following is the client code:

```
/*
  PUB - SUB envelop messages.
  client.c
*/

#include "czmq.h"


int main (int argc, char const *argv[]) {

  zctx_t* context = zctx_new();
  void* request = zsocket_new(context, ZMQ_SUB);

  printf("Starting client...\n");

  zsocket_connect(request, "tcp://localhost:4040");
  zsocket_set_subscribe (request, "Company10");

  for(;;) {


    char* env = zstr_recv(request);
    char* msg = zstr_recv(request);
    printf("env: %s | %s\n", env, msg);

    if(!msg) break;
```

```
        free(env);
        free(msg);

        zclock_sleep(1);
      }

      zsocket_destroy(context, request);
      zctx_destroy(&context);

      return 0;

    }
```

High watermark

When messages are sent from source node to destination node very rapidly, the source node could run out of memory. There are quite a few solutions to this, such as flow management. However, it may not be feasible in some situations.

ZeroMQ uses a high watermark to set the capacity of pipes. In ZeroMQ v2.*x*, the default value is infinite whereas in ZeroMQ v3.*x* the default value is 1,000.

When a socket reaches a high watermark, it either drops the message or blocks it. The REQ-REP socket will block the message whereas PUB will drop it.

The following is an example:

```
/*
  HWM example.
*/

#include "czmq.h"


int main (int argc, char const *argv[]) {

  zctx_t* context = zctx_new();
  void* pub = zsocket_new(context, ZMQ_PUB);
  zsocket_bind(pub, "tcp://*:4040");

  zsocket_set_hwm(pub, 10);
```

```
    for(;;) {
      zstr_sendm(pub, "Company1");
      zstr_send(pub, "Message to be ignored.");
      zstr_sendm(pub, "Company10");
      zstr_send(pub, "Message to receive.");
      zclock_sleep(10);
    }

    zsocket_destroy(context, pub);
    zctx_destroy(&context);

    return 0;
  }
```

Reliability

Applications may crash unexpectedly, stop responding, can have memory leaks, or a bug can make them run slower. In addition to problems that an application may have, we may experience hardware failures or network problems. We need to be sure that messages arrive at their destination no matter what problems our infrastructure may experience. Reliability means every event is guaranteed to arrive at its destination.

Most message queue implementations rely on a broker to have reliability, which means messages are queued and then delivered to their destinations, whereas in ZeroMQ, applications directly communicate with each other and messages are resent if they are lost for some reason.

It is easy to figure out if either the server or the client stops responding when we use the request-reply pattern. If the client or the server does not receive messages from each other, it means there is a problem.

If you recall from *Chapter 2, Introduction to Sockets*, we said that the publisher does not know whether a subscriber is connected or not. This also means that if a subscriber starts experiencing problems, the publisher will not know about it and the subscriber will miss messages the publisher has been transmitting.

When it comes to reliability in the publish-subscribe pattern, we need bidirectional communication between the publisher and the subscribers. However, the publisher-subscriber pattern does not support bidirectional communication in ZeroMQ, therefore, the option is to use the dealer-router pattern.

Having reliability in the request-reply pattern is relatively easier than in the publish-subscribe pattern. We could simply retry sending the message if we have not received a reply yet. If we still do not get a reply after trying a number of times, we could discard the communication.

Heartbeating is a layer that can be used to detect if a worker has died or is alive. However, it should not be used with the request-reply pattern. Heartbeating travels asynchronously between resources.

If there are a limited number of subscribers connected to the publisher then TCP is fine, whereas if there are massive number of subscribers, in that case, it would be a better idea to use PGM.

Slow subscribers in a publish-subscribe pattern

A serious issue of the publish-subscribe pattern is slow subscribers. A flawless environment would be one where the publisher sends messages to the subscriber at full speed, but this is utopia. In reality, subscribers cannot keep up with the publisher most of the time. They are either poorly implemented, have network issues, or some other reason.

Let's consider the following example where the subscriber runs slower and we abort the program. First, let's look at the server code:

```
#include <stdio.h>
#include <string.h>
#include <unistd.h>
#include <sys/time.h>
#include <time.h>
#include "zmq.h"

int main (int argc, char const *argv[]) {
  void* context = zmq_ctx_new();
  void* publisher = zmq_socket(context, ZMQ_PUB);
  printf("Starting Server...\n");

  zmq_bind(publisher, "tcp://*:4040");

  for(;;) {
    time_t current_time = time(NULL) % 86400;
    char str[11];
    snprintf(str, sizeof str, "%lu", current_time);

    int s_len = strlen(str);
```

```
    zmq_msg_t message;
    zmq_msg_init_size(&message, s_len);
    memcpy(zmq_msg_data(&message), str, s_len);
    zmq_msg_send(&message, publisher, 0);
    zmq_msg_close(&message);
    sleep(1);
  }
  zmq_close(publisher);
  zmq_ctx_destroy(context);

  return 0;
}
```

And the subscriber that runs slower:

```
#include <stdio.h>
#include <stdlib.h>
#include <string.h>
#include <unistd.h>
#include <sys/time.h>
#include <time.h>
#include "zmq.h"

#define DELAY 4

int main (int argc, char const *argv[]) {

  void* context = zmq_ctx_new();
  void* subscriber = zmq_socket(context, ZMQ_SUB);

  printf("Getting data...\n");

  int conn = zmq_connect(subscriber, "tcp://localhost:4040");
  conn = zmq_setsockopt(subscriber, ZMQ_SUBSCRIBE, 0, 0);

  int i ;

  for(i = 0; i <  10; i++) {
    time_t current_time = time(NULL) % 86400;

    zmq_msg_t reply;
    zmq_msg_init(&reply);
    zmq_msg_recv(&reply, subscriber, 0);
```

```
        int length = zmq_msg_size(&reply);

        char* value = malloc(length + 1);
        memcpy(value, zmq_msg_data(&reply), length);
        zmq_msg_close(&reply);

        unsigned long t_timer;
        sscanf(value, "%lu", &t_timer);

        int res = abs(current_time - t_timer);

        free(value);

        if(res > DELAY) {
          printf("Subscriber is too slow. Aborting.\n");
          break;
        }
        sleep(3);
      }

      zmq_close(subscriber);
      zmq_ctx_destroy(context);


      return 0;
    }
```

We have defined a `DELAY` constant in our subscriber code and we calculate the time the server takes to send a message and the local time of the subscriber. If the difference between these values is larger than the `DELAY` constant, we abort the subscriber as it means it runs slower. This is also known as **suicidal snail pattern** in ZeroMQ terminology.

Summary

In this chapter we briefly looked at some advanced patterns, reliability, and how to deal with slow subscribers in the publish-subscribe pattern.

ZeroMQ is a flexible, easy-to-use, and fast message queuing service. Unlike other message queuing libraries, it allows developers to implement their own message queuing services.

Appendix

Structure and Interpretation of Computer Programs, Second Edition, Abelson Harold, Gerald Jay Sussman, and *Julie Sussman, McGraw-Hill Science,* 1996.

Fuzz revisited: A re-examination of the reliability of Unix utilities and services, Barton P. Miller, David Koski, Cjin Pheow, Lee Vivekananda Maganty, Ravi Murthy, Ajitkumar Natarajan, and *Jeff Steidl,* 1995.

CSSV: Towards a Realistic Tool for Statically Detecting All Buffer Overflows In C, Nurit Dor, Michael Rodeh, and *Mooly Sagiv,* 2003.

Beej's Guide to Unix IPC, Brian "Beej Jorgensen" Hall, available at `http://beej.us/guide/bgipc/output/html/multipage/index.html` and viewed on October 22, 2012.

ZeroMQ – The Guide, Pieter Hintjens, available at `http://zguide.zeromq.org/` and viewed on November 4, 2012.

ZeroMQ – API, Pieter Hintjens, available at `http://api.zeromq.org/` and viewed on October 17, 2012.

CZMQ – High Level C Binding for ZeroMQ, Pieter Hintjens, available at `http://czmq.zeromq.org/` and viewed on November 6, 2012.

A TCP Tutorial, Andy Ogielski. available at `http://www.ssfnet.org/Exchange/tcp/tcpTutorialNotes.html` and viewed on October 15, 2012.

An Introduction to Asyncronous Programming and Twisted, Dave Peticolas, 2011, available at `http://krondo.com/blog/?page_id=1327` and viewed on September 12, 2012.

ZeroMQ an Introduction, Nicholas Piël, 2010, available at `http://nichol.as/zeromq-an-introduction`.

Valgrind Manual, Julian Seward, available at http://valgrind.org/docs/manual/quick-start.html#quick-start.intro and viewed on October 17, 2012.

Cloud-based Queuing System with Strong Consistency, Zhe Zhang, Han Chen, Minkyong Kim, and *Hui Lei,* 2011.

Message Queue Evaluation Notes, Second Life, available at http://wiki.secondlife.com/wiki/Message_Queue_Evaluation_Notes#Zero_MQ and viewed on September 15, 2012.

Index

[PACKT] open source
community experience distilled
PUBLISHING

Thank you for buying
ZeroMQ

About Packt Publishing

Packt, pronounced 'packed', published its first book "*Mastering phpMyAdmin for Effective MySQL Management*" in April 2004 and subsequently continued to specialize in publishing highly focused books on specific technologies and solutions.

Our books and publications share the experiences of your fellow IT professionals in adapting and customizing today's systems, applications, and frameworks. Our solution based books give you the knowledge and power to customize the software and technologies you're using to get the job done. Packt books are more specific and less general than the IT books you have seen in the past. Our unique business model allows us to bring you more focused information, giving you more of what you need to know, and less of what you don't.

Packt is a modern, yet unique publishing company, which focuses on producing quality, cutting-edge books for communities of developers, administrators, and newbies alike. For more information, please visit our website: `www.packtpub.com`.

About Packt Open Source

In 2010, Packt launched two new brands, Packt Open Source and Packt Enterprise, in order to continue its focus on specialization. This book is part of the Packt Open Source brand, home to books published on software built around Open Source licences, and offering information to anybody from advanced developers to budding web designers. The Open Source brand also runs Packt's Open Source Royalty Scheme, by which Packt gives a royalty to each Open Source project about whose software a book is sold.

Writing for Packt

We welcome all inquiries from people who are interested in authoring. Book proposals should be sent to author@packtpub.com. If your book idea is still at an early stage and you would like to discuss it first before writing a formal book proposal, contact us; one of our commissioning editors will get in touch with you.

We're not just looking for published authors; if you have strong technical skills but no writing experience, our experienced editors can help you develop a writing career, or simply get some additional reward for your expertise.

[PACKT] open source
community experience distilled
PUBLISHING

Mastering OpenLDAP: Configuring, Securing and Integrating Directory Services

ISBN: 978-1-847191-02-1 Paperback: 484 pages

Configuring, Securing, and Integrating Directory Services

1. Up-to-date with the latest OpenLDAP release
2. Installing and configuring the OpenLDAP server
3. Synchronizing multiple OpenLDAP servers over the network
4. Creating custom LDAP schemas to model your own information

---

Catalyst

ISBN: 978-1-847190-95-6 Paperback: 200 pages

Design, develop, test, and deploy applications with the open-source Catalyst MVC framework

1. Understand the Catalyst Framework and MVC architecture
2. Build and test a site with Catalyst
3. Detailed walkthroughs to create sample applications
4. Extend Catalyst through plug-ins

Please check **www.PacktPub.com** for information on our titles

CPSIA information can be obtained
at www.ICGtesting.com
Printed in the USA
FFOW01n1323050414
4708FF

9 781782 161042